My best wishes on your [illegible].
May life's cycles find you both
always ready to adapt to changes
and new challenges.
Fondly
Eva

Tideline

Tideline

Ernest Braun

A Studio Book • The Viking Press • New York

To the limpets, barnacles, birds, and
the whole community of tideline life

First published in 1975 by The Viking Press, Inc.
625 Madison Avenue, New York 10022

Published simultaneously in Canada by
The Macmillan Company of Canada Limited

LIBRARY OF CONGRESS CATALOGING IN PUBLICATION DATA

Braun, Ernest.
Tideline.
(A Studio book)
Bibliography: p.
1. Tide pool ecology. 2. Seashore ecology.
I. Title.
QH41.5.S35B7 574.5′2636 75-12719
ISBN 0-670-71255-8

Text printed in U.S.A.
Photographs printed in Japan

Second printing May 1976

Contents

Introduction

Tideline is an account of one day's change of ocean tides at the meeting place of land, sea, and air. Starting with high tide at dawn, this report of what I observed follows the waterline as it recedes from cliffs, across sand, through rocks and pools to the midday low, and returns with the flood tide. The ocean makes this short journey up and down the beach, twice a day, to dominate a thin band of wilderness reaching for thousands of miles around earth's landmasses.

If we look closely enough to see the mysterious and intense patterns of beauty and try to understand the success and diversity of life, the time spent moving with the tide across this narrow wilderness can be the beginning of countless journeys. Four thousand waves will break on this beach during the tidal change, expending energy from sea winds and distant storms. The long evolutionary journey of life is well documented here by a host of plants and animals precisely designed for survival in this harsh environment. The rhythms of the earth's rotation, its travels around the sun, and the moon orbiting earth control the tidal pulse. To understand truly what is here would mean exploring all the natural sciences and then making trips into space and back through time to fill in the gaps.

Although my day spent at the tideline is crowded with impressions and discoveries, it is only an introduction to this zone claimed by both land and sea. And yet it is a day different from all other days, for this combination of tide, weather, season, and time of day has never occurred before and will never happen again. A comprehensive statement on the tideline would fill many volumes and consume more than one lifetime. This book is a progress report, limited by my state of awareness.

My first photographs at the beach were made during family vacations. As I became more involved and moved closer with the camera, I tried to discover and communicate the beauty and constant change of the tideline. I soon became curious about the strangely beautiful plants and animals that thrive in extremely changeable conditions. I have been surprised and delighted to find a vast resource of written material with answers to all my obvious questions. Many of the available books convey feelings of beauty and aliveness as well as the factual material that I sought. The words and pictures in this book will serve a good purpose if they stimulate the reader-viewer to use the bibliography.

A pall seen and unseen now hangs over the tideline and all wilderness. On the beach I see oil fouling tide pools and plastic containers littering the sand, but I cannot see the increasing radioactivity in ocean water or the depletion of schools of haddock over Georges Bank. These ominous signs of man's arrogant mastery of earth's environment must be acknowledged and evaluated. Can man once again live in harmony with the natural world? Our survival as a family of living things on this planet depends on how we answer this question.

High Tide

The crash of wave against rock, the roar and clatter of water receding down rocky beaches: all the clamor of the rising tide is sweeping through cliff-top forests and meadows with the on-shore winds of a growing storm. I hear the steady pulse of sea beating against land as it has since the first rains filled the ocean basins. It beats loud and strong as the moon and the sun pull together at the spinning earth to make the highest tide of the month.

In this hour before dawn, I again follow the trail to the sea, along this spiny point, deep into groves of pine and cypress. Only vague shadowy shapes are visible, leafy arms swaying in a blue-black sky. Ancient trunks creak, branches rattle, thick cypress crowns wail and shriek at the rising gale. Rain has a way of releasing the good earthy smells of the forest. Damp leaves and pine needles, spicy sage mixed with the heavy woodsy aroma of pine and cypress blend with that indescribable smell of the sea. Salt spray mixed with rain is flying with the wind. I can taste it—bitter, tangy, laced with memories of sparkling sun on water, the joy of discovery, good times on many beaches. As the wind and salt air clear the sleep from my head, I feel a growing eagerness for whatever the day will bring.

As I hurry along, my thoughts return to a beach I loved as a small boy. I feel the same eagerness now, the same tingle of anticipation of the always new, strange, and very beautiful things to see and to find. At low tide my friends and I would race the waves around the point to a reef where I found my first starfish. Sometimes there were sharks' eggs, and jellyfish that hid under a cloud of purple ink when they were poked. I scrambled after the wily shore crabs. There were shells to collect and long strands of giant kelp to marvel at. Now, thinking back, I feel that the sea has always had a deeper, primitive attraction for me, for all of us. Perhaps it is because sea water and human blood are so similar. We broke our living bond with the ocean in the dim shadows of the past, when life first moved from the sea to colonize the land. But some inside part of each of us knows that the beginning was in the sea.

As I approach the cliff top, dark clouds hurry across the sky, hiding the glow of pink that had been growing in the east. From the sea great dark masses rush from the gloom to smash at rocks and cliffs. I feel the land shake and hear the boom and the thundering roar but can see nothing except wind-driven spray and rain.

Then, in brightening dawn, I see that the beach is covered with seething gray and white foam reaching for the cliff tops. Waves, ten or twelve feet high, surge out of the mist in endless ranks, plumed with gale-driven spume. The largest waves bound past all rocky barriers to explode against the land with mighty bursts of foam and spray.

At the base of the cliff, families of limpets tighten their grip on rocky walls with suction-cup muscles, while leaf barnacles resolutely turn their armored plates into the onslaught. Beneath the waves hermit crabs scurry for the shelter of cracks and crevices, as turbulent currents move seaward along the bottom scattering debris and pebbles along the roughly sloping shore. On the outer edge of the rocky beach, tough, rubbery sea-palm algae bend under tons of white water and foam. More than one of these outpost guards will be left high on the beach among the cast-offs of this storm, their holdfasts torn loose from the outer reef along with whole communities of tiny creatures that live in these rootlike forms. Even the gulls have taken shelter in ponds and meadows behind the beach. The only living creatures visible now below the cliffs are a few hardy cormorants, calmly waiting for breakfast on the tip of an offshore rock just above the reach of the highest waves. One Brown Pelican skims along a wave trough, almost touching the water. His wings move just three times before he merges with the morning mist of spray and rain.

In this glorious moment all the forces of earth, moon, and sun release their energy, aided by the full fury of wind and rain. The power of the sea has reached its height. I have never known a dawn as alive as this. There will never be another just like it. As the cliff shakes beneath my boots, I am reminded that the sea may claim this spot at any moment in the next billion years for its domain, which already covers three quarters of the earth's surface. Yet the land, I know, will defend its boundaries at its own majestic pace. This western edge of the continent that supports me is rising, only a few inches in my lifetime, but by a movement of leaps and bounds in geologic time. This patient land has learned to use the moving sea in its own defense, trading a small section of high cliff for a long sloping beach, donating some beach sand to parallel currents that build a bar to protect a cove between two rocky points. Over the seasons the cliff will regroup its materials to straighten the perimeter against this continual attack.

Now I can see the meadow behind the cliff. Orange-petaled poppies that reached for the sun yesterday have furled their blossoms and are hidden under wind-flattened grasses. As I turn to watch the cliff-hanging cypresses break the full force of the gale for the first time in six thousand miles of open sea, a wave breaks high over the top of the cliff. Above me a solid sheet of spray laced with seaweed hangs for a long moment, then drops, leaving me drenched and frightened.

There had been no warning that this giant was coming. Of all the eight thousand waves that release their energy against this shore during an average day, why was this monster the largest? Where had it come from?

Crouching in the lee of a cypress I search my memory for clues to piece together my own version of the long journey that has just ended. It is possible that this mighty, wind-formed wave has carried energy from a monsoon off the China coast to batter the outer wall of this continent below me. Four and a half days earlier, on the surface of a warm calm sea, the wave was born. . . .

Far to the southwest of Alaska, where the Kuroshio current moves north past Japan on its way to the Arctic, the wave's birth is assisted by some of the most powerful forces in nature. Waves owe their existence to massive movements of water and air, changes in temperature and pressure on a global scale, forces generated by the rotation of the planet on its axis and by the energy of the sun that strikes the orbiting earth.

Following the tendency for all moving things—wind, current, or bullet—to turn to the right in the northern hemisphere and to the left below the equator, the Kuroshio moves in an often interrupted but clockwise path around the North Pacific. This is one of the great planetary currents that pushes sun-warmed water from the equator toward the polar ice caps, recycling ice-cooled water as the circle is completed. These great rivers in all the oceans are as constant as the rotating earth that forms them, changing only as continents rise and fall over millions and billions of years. The constant circulation of sea water helps to equalize the temperature of landmasses, modify climate and weather. Companions of the currents, rivers of air, the tradewinds—the Westerlies, the Easterlies—blow consistently across the sea in certain latitudes. Navigators of ships using wind power and sails long ago learned to rely on these great air streams.

The wave that had just drenched me had its beginning between 20° and 30° N latitude and 140° longitude. On the first morning the warm current moves across a calm sea, reflecting a clear sky. Water and air here have the same temperature at their interface. This peaceful balance begins to change as radiant heat from the sun, now high in the sky, penetrates the flat sea at its most effective angle to raise the temperature on the surface. As the current carries the warming sea steadily northward toward colder regions, heat is gradually absorbed by the air next to the warm surface. This warmer air expands, grows lighter, then rises, thus lowering the air pressure. Immediately, cooler air at higher pressure begins to move in to balance the atmosphere.

At the moment of birth this first puff of wind barely ripples the mirror surface. All afternoon a freshening breeze forms ripples and then wavelets, scattering an occasional whitecap over the choppy blue-green ocean. As darkness approaches, long bands of cirrus clouds color the spray from six-foot seas with their own fiery glow. A fresh southeast breeze of twenty-one knots pushes the seas into higher and steeper waves, which continue to grow and absorb energy from the wind, flowing freely for hundreds of miles across the open sea. After dark the wind holds steady at twenty-five knots; but the series of waves—the train, as it is called—continues to grow in height, length, and period (the time in seconds for a wave crest to travel a distance equal to one wave length). Finally, less than twelve hours after birth, the wave is full-grown. It has absorbed all the energy from the wind that is possible at that velocity.

Each wave is different from all others. They constantly meet and exchange energy with other waves of all sizes moving in many directions. The wave we follow is overtaking smaller waves and adding their energy to its own strength. Racing through the night toward the China Sea, the wave moves on the surface as sound, light, and radio waves move through the air. It is pure energy, moving without transporting air or water. Particles of water in the path of the wave move in a circular orbit equal to the height of the wave until each particle returns almost to its original position. Since virtually no water moves along with the wave, only a small amount of energy is needed to power these orbiting particles and move the wave.

In the early light of the next gray, blustery dawn, the wave drives through swells from a growing disturbance in the Philippine Sea. Above its white crest the dark shape of an albatross passes over the turbulent sea. It flies northward, away from the monsoon. Its great dark wings beat against the rising gale, for there is no respite in the scene below. Rolling, pitching mounds of salt water churn the surface. Streaks of spindrift fly before the wind. Plunging through this violent sea, the wave now smashes into a larger wave from the gathering storm; using most of its energy to slow the speed and diminish the height of this giant wave, it squanders what is left in a cloud of gale-blown foam.

Now the wave born in the China Sea is absorbed by a forty-five-foot storm giant rushing eastward on an opposite course. The former ripple keeps growing as it gathers the energy of other waves in the circular path of the monsoon. Now it towers far above its fellows and is exposed to the full fury of the seventy-knot hurricane. The unstable crest is torn away by a screaming gust which hurls the huge mass into the trough between waves to form streamers of driving spray. The weight of this falling water, moving independently of the wave, could do great damage to anything afloat.

The storm continues to nourish the wave during a twenty-five-hundred-mile journey to the northwest before dissipating its circular canopy of dark clouds, rain, wind, and high seas. As the monsoon dies it transfers most of its energy to a series of large swells that will surge past the Bering Sea with enough energy to keep moving across thousands of miles of open sea. Encouraged by the North Pacific Drift, the great wave, now part of this new train, passes south of the Aleutians.

For two more days this pulse of energy journeys over a sea of many colors: blues, purples, greens, grays, and the reds of dusk. It lifts the fishing fleets of three nations and gently rocks a pair of sleeping Arctic terns. Some nights the wave moves over glowing billions of phosphorescent plankton. Here the surface is pierced by flashes and streamers of light, as the tiny plankton eaters attract hunters who themselves become food for larger fish.

Moving now through the California Current south of the Alaskan Gulf, the wave has dwindled to one third of its former size, yet it is still strong enough to push on through wave trains from the Arctic storms that come howling south toward the western edge of North America. Absorbing the energy of swells in its path and gaining vigor from the winds of a series of spring squalls, the wave hurries toward land on its last night.

Running before the wind on a diagonal approach to the coast, the wave crosses the continental shelf and in less than a minute begins to drag on the bottom. As the wave's water particles hit the sloping shelf their circular orbit slows. The leading section of the wave loses speed until the whole mass turns to move parallel to the slope of the shore. Other waves, coming from different directions and distances and having different heights and periods, also turn to form orderly ranks as they too approach the shore.

As this great surge of energy rushes into shallow water, its sides become steeper and its base narrower. Now there is not room under the wave for water particles to complete a circular orbit and the crest "peaks up." Because the tide is now at its height, the immense wave hurdles the outer reef to slam onto the cobblestone beach, its gray-green body translucent along the steep crest. Near the cliff the peak of the wave is moving faster than the base. So, following its orbital shape the crest hammers the cliff, releasing in an instant all the power that carried it across a great expanse of ocean through storm and calm. Green water joins with air in a final explosion of white foam to mark this abrupt end to a long journey. . . .

Again this cliff has stood firm against one more wind-spawned surge of energy in the endless series that forms the pulse of the living ocean. While I, cold and wet thanks to rain and wave, with my poncho billowing out behind me like a sail, concede that the tide and the storm are in control.

The tireless demolition team of the sea does its best work when the tide is high. Water moving at great speed can trap and compress air with enough pressure to shatter rock. Storm waves can pick up pebbles and rock fragments that bombard the cliffs. Sand in suspension forms a cutting edge to undercut the cliff wall. Rocks of all sizes grind away at each other. Boulders break up into pebbles and pebbles become sand, the ultimate product of the erosion cycle on the beach.

On this beach that I know so well, the cliffs that form a continuous backdrop for the tidal zone can be read like a long page of world history. Secrets usually hidden under the surface are revealed: how the land was formed, its age, a record of rising and sinking landmasses. Layer after layer of sedimentary clay and sandstone were obviously built up from the floor of the sea to be lifted and tilted into the cliff wall. Fossil shells in some layers confirm this origin. In places the ocher-colored stratum is marbled with purple, brown, and red pebbles the same size and color as the loose pebbles on the beach.

Sea sculpture, the incredible shapes formed in land by sea and wind, make up the great collection of three-dimensional art we call the beach. This exhibit is always changing. There are beaches that I visit again and again with the feeling that I am in an artist's studio. It seems disrespectful to tramp over the beautifully carved layers of stone along the arc exposed by a receding tide. Here the tools and the finished art are equal in beauty. The tools, chunks of igneous rock smoothed into cobbles and pebbles over the eons, grind away at the softer layers of more or less horizontal sedimentary formations, using the energy of moving water. Changes in level are starting places for this carving. Gravity and turbulent water continue the process. In some holes the same pebbles may have been grinding and smoothing the same shape since Columbus demonstrated that the earth is not flat.

The sound of waves moving over a steep beach of pebbles or cobbles makes my spine tingle. This chorus of clinking surfaces is penetrating and satisfying in an indescribably delicious way. Each pebble, so uniquely shaped, has always seemed a special treasure to me, especially when it is wet and glistening. Why, I wonder, are no two pebbles exactly alike? How long does it take for a pebble to become a grain of sand?

The cycle of earth's daily rotation from west to east has now placed the beach well within reach of sunlight. As the storm slackens I am suddenly aware of being wet and cold. Visions of a warm fire and that first steaming mug of morning coffee are rapidly becoming more important to me than the drama of dawn on this wild beach. It's time to start back through the cliff-top forest to shelter and warmth, which seem to be a human necessity.

The cliff-dwelling trees rooted in rock, their limbs washed by salt spray and shaped by sea winds, are a most graphic example of the intimacy of land and sea. These trees would not be here were it not for the moisture of flying spray, fog drip, and the year-round air conditioning of relatively stable, cool ocean temperatures. The prevailing winds are quite precisely graphed by the growth patterns of the most exposed trees. The colony of cypress that cling to this rocky point can live for two or three hundred years. The silvery, bleached skeletons of the pioneer trees record their long struggle to reach for the sun while grasping the rocky cliff strongly enough to withstand the gales of winter. Resilience and determination show in exposed root systems. Trunks are buttressed and shaped to support gale-blown branches.

As if to complement the mosaic of rocks, pebbles, and sand on the beach below, a rich tapestry of growth spreads over cliffs and rocks in the shade and shelter of cypress and pine. Against a background of green and gray trees grow the blue and purple lilac, wild iris, and cliff daisies, the orange and red of sea fig and paintbrush. This is the spectrum of spring. In midsummer cliff succulents will have yellow blossoms atop long, red stems and the harvest colors of flowering buckwheat and dry grasses will be added. This whole forest community—the insects, birds, squirrels, and deer—is directly dependent on moisture from the sea. Although this forest receives some winter rain, it is not nearly enough to support such abundant and varied life. On the eastern slope of the coast range only a few miles away, the long days of summer sun will soon leave the land baked and dry. Only plants with roots reaching deep below the parched surface will stay green until the winter rains.

The prevailing wind here, as on most coasts, blows from the sea. This is particularly true in the summer months when living things must depend entirely on these cool, wet breezes. Typically, the early morning sun warms the land near the sea faster than it does the ocean surface. This warm land begins to radiate heat into the air, which then expands and rises, lowering the air pressure. Cool, moist offshore air, which is denser, moves in to equalize the air pressure. The planetary current circling the North Pacific brings cold water from Alaska past the Northern California coast all summer. Since here the coast angles out into the current, and for other more complex reasons, colder water from the depths of the sea is circulated to the surface to lower the temperature even further. Fog forms as moisture-filled air cools enough for water vapor to condense into visible droplets. Heat from the sun causes water to evaporate continually from the ocean surface. At dusk, when this heat source is gone, fog will form before sunset. Tiny wisps become long tendrils until, in a few moments, the sky that was blue and gold is gray. This evaporation cycle on a global scale circulates water vapor from sea to clouds, as rain from clouds to land, and then returns it to the sea through rivers and streams.

From the tip of this steep granite headland, the forest trail contours through pine and cypress past a series of small coves, past offshore islands, outposts of jagged rock, shouldering the great ocean swells away from intimate, cliff-guarded beaches. The thick, damp carpet of needles underfoot is segmented almost geometrically by a web of shallow pine roots forced to the surface by the sparse topsoil on this granite rock. Where one of the small cones has lodged on the wrinkled cliff face or been scattered in a handful of soil by a ground squirrel, pines, pursuing their dwarfed and precarious lives, push upward beside the lichens and cliff lettuce.

Gray-green fronds of lace lichen decorate all the trees that have even a hint of protection from direct sun or wind. This lichen is part fungi, trapping moisture from sea air, and part algae, which make food from air, water, and sunlight by the intricate chemical process of photosynthesis. This friendly and mutually beneficial relationship between living things is called symbiosis. The lace lichen is a good neighbor that makes no demands on its tree host other than living space. This morning the tree lichens are streaming out like gauzy flags before the southwest wind.

The roaring, howling, sighing voices of many trees protesting the gale are suddenly hushed as I reach the protected heart of this ancient cypress grove. It is like walking into a musty old building, abandoned now but obviously solidly constructed by master craftsmen. The dense, wind-flattened crowns of these venerable trees form an almost solid roof that leaks in only a few places. Dull gray trunks are pillars, leaning and warped with age and abuse. The lichens that cobweb the old structure are set in motion by the same kind of damp, cold draughts that come through broken windows and doors. Like scaling paint, the amazing red algae, found only in very damp and shady places on the cypresses, cover branches and twigs. These algae make a living in the same way as the lace lichen. Their color comes from pigment in the cell sap.

Hurried along by thoughts of my morning coffee, I become aware of a persistent cracking sound. A sea otter and her baby have come close to shore for breakfast in this protected cove. The mother, shaped like a small seal with paws instead of flippers and with a most intelligent face, floats on her back while pounding a sea urchin against a rock held on her flat stomach. These loud preparations are followed with great interest by the little one, who moves about its mother and finally in a graceful movement reaches up to her mouth for its share of the meal. When diving for food, the sea otter always brings up the right tool with which to crack the urchin. Farther from shore in the thick kelp, two groups or "rafts," as they are called, of sea otters are quite at home dozing and preening. Since otters lack the insulating blubber of other sea mammals, a layer of air bubbles trapped in their soft fur maintains buoyancy and protects the warm-blooded creatures from frigid waters. The sleeping animals use streamers of kelp as sea anchors for their water beds. With paws clasped together they float motionless in sleep, rocked by the gentle swell. These mammals of the tideline never venture onto the shore. They were practically exterminated by fur hunters and so have good reason to distrust creatures of the land. In 1938 they were seen again and now are one of the joys of cliff-watching on sections of the Northern California coast and in some parts of the Aleutian Islands. When the tide is high, the otters move in to feed in a tidal zone that would be dry at any other time.

The earth's rotation cycle now reveals the sun halfway from the eastern horizon to the high-noon midpoint. Three hours have passed since dawn and the apex of the spring tide. The earth has completed one sixth of its daily rotation.

The midmorning sun now pokes bright blue holes in the overcast as I return to the beach. A crackling fire and that steaming coffee have accomplished for me what the sun is doing now for the bedraggled poppies along the trail. The colors of spring have replaced the blacks, browns, and grays of dawn. Vivid green grass dotted with poppies in the cliff-top meadow beyond the forest leans before the stiff sea breeze. Blue water with hints of green and purple veined with white, playing with offshore rocks and the reef, shows little of its early morning hostility. On the beach a great change is taking place. The largest waves no longer break directly against the cliff. The rocky brow of an offshore reef now challenges the gray-green swells. Below me at the base of the cliff a strip of pebbles and sand widens. On rocks that were only splashed during the highest tide, limpets have finished grazing on algae and return to their low-tide homes, while each of a host of barnacles prepares for the long, dry hours ahead by closing tight its four armored doors to retain some life-giving water. Every living thing in the crowded tideline community is profoundly affected by this change in water level. Surrounded by the reality of this changing beach, the curious part of me tries to understand the vast rhythms of rising and falling tides.

Waves as long as half the distance around the earth form the tides. The crest and trough of the wave are known as the high and the low tide. These waves have a period of twelve hours and twenty-five minutes. This is the time interval for each succeeding wave crest or high tide to pass a given point on a beach. The height of the tide, or of the wave, is measured only near the shore, as it is greatly influenced by the shape of the shore.

This long wave or tidal bulge is caused by the gravitational pull of the moon on earth. While the earth rotates on its axis once every twenty-four hours, the tidal bulge moves with the moon on a twenty-eight-day cycle of rotation around the earth.

The gravitational pull of the sun also influences the tides. Add the sun to your mental picture of earth and moon. It has a mass twenty-seven million times greater than the moon; the gravitational attraction of the earth for the sun is one hundred fifty times as strong as the earth's attraction for the moon. The moon, however, is a great deal closer to the earth and the distance factor is decisive. The water particles on the side of the earth facing the moon are pulled more strongly than the particles on the side facing away from the moon. This difference in attraction from one side of the earth to the other causes the bulge. The sun is so much farther away than the moon that the difference in its gravitational pull from the two sides is one half as strong.

The centrifugal effect of the rotating earth-moon system forms an equal bulge of water particles on the side of the earth opposite the moon. This is why there are two high and two low tides each day.

The moon revolves in the same direction the earth is spinning, so that any point on earth will have to go slightly farther than one rotation to come directly under the moon again. This is why the tidal day is twenty-four hours plus fifty minutes instead of twenty-four hours. The rough-bottomed ocean basins tend to drag the tidal bulge along the path of rotation, so that the crest of the wave or the high tide is slightly ahead of the moon instead of directly under it.

The tides are a great example of the sensitivity of liquids to gravity, centrifugal force, and other natural laws. To understand these natural laws and how they affect the sea is a lesson in awareness. I have been back many times to the encyclopedia to find out a little more about gravity and am reminded that each particle or mass is attracted to every other with a measurable force. In reality, then, every particle of water feels the gravitational pull not only of the moon and the sun, but of every star in the universe, and, though we are not aware of it, we too are affected by these same forces.

The influence of the sun on the tides is important because it does increase or decrease the lunar tides. When the pull of the sun and the pull of the moon are aligned in relation to the earth, the solar bulge is added to the lunar bulge to increase both the high and the low tide. This is called the spring or in-phase tide. When the sun and the moon are not aligned, the solar bulge decreases the lunar tide. This is called the neap or out-of-phase tide. Between these two extremes the solar bulge warps the main bulge, and high and low tides come a little earlier or later, slightly changing the length of the tidal day.

The elliptical orbit of the moon creates another variable. When the moon is at its closest to the earth, at the perigee of the orbit, the moon is fifteen thousand miles closer than at its farthest point, and the tides are then twenty per cent higher and lower than normal. When the perigee of the moon and the sun are in phase—and when the moon is full or new—the highest tides of the year occur. This happens about twice a year.

The shape of the ocean basin affects the height and character of the tide. No tidal measurements have been made in mid-ocean, but we might assume that the range is only a foot or so, as it is on the shores of small mid-ocean islands. When the earth turns under the tidal bulge, the continental shelf acts like a wedge driven under the bulge. Openings in the shoreline, such as bays and estuaries, tend to amplify the wave even more.

One of the many other complex factors causing variation in the tides is the oscillation of any body of water. Disturb a container of water and it moves with a rocking motion. The water at the edge of the basin moves more than the water in the center. In the exact center, called the node, there is no movement. The length and depth of the basin determine the period of oscillation. In the oceans the rotating earth under the tidal bulge keeps the oscillation moving. This can result in tidal variations of a foot or less at the node to forty feet within a few hundred miles of the center.

Even though the water level may rise or fall several feet in a few hours, it is not easy to see the tide actually change. I always check a tide table before spending time at the tideline. Beaches with steep cliffs may be easy to reach at low tide, but can be hard to leave when the tide is high. In *Between Pacific Tides,* Ed Ricketts, who spent many years collecting and observing tide-pool life, cautions that an unexpected high wave is a threat whenever one ventures below twenty feet above the mean tidal level. He suggests that it might be healthier to cling like a limpet rather than run or climb, if the need should arise. The chance of encountering a huge wave would be greater on an exposed beach during or between winter storms.

I learned the virtue of keeping one eye on the surf while working with camera and tripod on rocks well above the breakers. While I was looking through the viewfinder, that once-in-a-while wave crashed against the rock, drenched me completely, and ruined my camera. After building a fire to dry myself, I went back up to the same spot, determined not to let a little water keep me from getting the picture—and the same thing happened again. This time I saw it coming and lifted the camera high, so while I was soaked twice I lost only one camera.

Now the strip of rocky beach at the cliff base is wide enough to expose some sand on the surf side. Descending quickly I encounter just above the high-tide mark a continuous band of black-stained rocks that are very slippery when wet. All over the world this band of black is the sign of the meeting of land and sea. It is most often composed of blue-green algae, the most primitive of green plants, containing only protoplasm and pigments. Besides producing oxygen, algae are one of the few life forms to convert nitrogen from the air into the nitrates required by all living cells.

Look a little closer at this dark zone of matted algae and chances are that you will meet the periwinkle. This little snail, in its dark, cone-shaped shell about one-half-inch long, has nearly made the transition from sea to land. Unlike the periwinkle of the low-tide zone, it has evolved a gill that works almost like a lung and allows it to breathe air. The periwinkle can endure many days without sea water, having adapted to the spring tides, which come only every two weeks, and now it cannot tolerate constant submersion.

The vegetarian periwinkle eats by scraping the rocks with a radula, a long ribbon carrying hundreds of rows of teeth. This ribbon, several times longer than the animal, is tightly coiled like a watch spring. It works like a file, except that when the teeth are worn down a new supply is moved up into place. These little files wear away the rock, grain by grain. One patient biologist has measured wear and tear on one tide pool for sixteen years. He has found that the rock level was lowered by three eighths of an inch.

Although most shellfish send larvae into the sea, the female periwinkle carries its eggs, each in a cocoon, until birth. The new little snail finds food within the cocoon, so that when it finally breaks through the egg sac and leaves its mother it is a complete little snail about as big as a grain of sand.

Sharp eyes are needed to see the little periwinkles because they seek protection from rushing water and bright sun in cracks and sheltered places. Because survival conditions are extremely difficult, the shape, color, and texture of living things here are very different from those of the more familiar plants and animals of the land. There is a great difference in scale between the periwinkle and the vast expanse of ocean, so the human eye five or six feet above rocks and sand must know what to expect or it will see only the postcard view.

I look forward to visiting with some citizens of this community as the tide recedes, leaving the lower beach open to dry-shod exploration.

While looking for periwinkles I see a flat, cone-shaped shell that at first appears to be part of the rock because of its similar color. It is the beautiful limpet which has evolved one of the simplest systems for living on wave-battered rocks. The wide, flat shell is perfect for deflecting water, and the large foot or muscle gives the limpet the mobility and the strength to cling tightly. Most marine biologists seem to agree that each limpet has a home place on the rocks where its shell fits perfectly. When waves beat against its home, the suction-cup muscle can hold snug and tight on all sides. As the tide ebbs, the shell makes a watertight bond with the rock, and the limpet sustains itself on moisture held in a groove just inside the shell.

When the tide is up, the limpet leaves home to feed on algae, quite like its neighbor the periwinkle. After about two and a half hours, appetite satisfied, it returns to its chosen spot. How does it know where home is? One experimenter made a deep groove across its homeward path. When the limpet reached the groove, it stopped and spent a long time seeming to decide which way to go. However, at the next high tide it moved around the groove back to its home. When one edge of a limpet's shell was changed with a file, it did not return home but moved about two feet from its home. On the fourth day it moved again. After eleven days it disappeared.

Limpets seem to favor living in clusters or villages that always fit perfectly into the setting, as if designed for purely aesthetic reasons.

There are, of course, many variations in size, color, and location of limpets. They have been classified, studied, and given Latin names that bridge language barriers but are difficult for the casual observer. Since this trip is limited to one change of the tide, we can only hope to make discoveries that may be studied in more detail at greater leisure.

It is hardly possible to miss the rough-textured patina of white acorn barnacles covering rocks that are most exposed to the direct action of heavy surf. Barnacles have adapted to the most difficult location in the whole beach profile. Imagine a landscape of millions of steep, white, cone-shaped hills, each no more than an inch high, formed of six armored plates interlocking to form a circle. The opening at the top of the cone can be tightly closed by a set of four plates that protect the master of this house from drying out at ebb tide.

When covered by water, the four doors swing open and, like a flag flying, a jaunty feathered plume rises from each hill in the rockscape. This plume is made of six pairs of thin, branched fibers that make an efficient net to strain the great sea soup for morsels of meat and vegetable.

Inside this fortress lives a pink shrimplike creature with its head firmly cemented to the floor of its house. It is the only member of the large class of animals called Crustacea that leads a sedentary life. Its mobility has been traded for a home with a strong foundation. If the barnacle is not caught with its front doors open by one of its enemies—a starfish, a worm, or a snail—it lives from three to five years. Its house then becomes a refuge for newborn periwinkles and other small citizens of the area.

The barnacle begins life in a cloud of milky larvae, hatched from its mother's shell into the waves. At first it is a little swimming creature called a nauplius, nourished by a glob of oil that not only feeds it but keeps it near the surface. As this buoyant food supply dwindles, the larva swims in lower levels until it finally reaches the bottom and is ready to become an adult. This larval stage lasts for three months. Now comes the challenging task of finding a home on a rock surface that is not too smooth or covered with algae. Once the choice is made, this little blob of jelly must withstand the full force of the sea while it builds its shell and cements itself firmly to the rock. This process takes only twelve hours. How many of these little ones find a safe spot and cling to it long enough to build the complex structure they need? It is thought that the larvae, now part of the plankton community, drift with the currents for many days, testing and examining many locations before making the final decision.

Once established in its new home, the young barnacle must change its clothes often, as its body outgrows its tough skin. How does it wriggle out of this tight unyielding skin with its head cemented to the floor? How does it keep enlarging the armored fortress as its body grows?

Barnacles often share their exposed position with the blue-black mussel. This shellfish has a conventional bivalve structure of two shells which is anchored securely to the rock with homemade mooring lines of natural silk spun by a gland in its foot. These lines stretch out in all directions but are more numerous in the direction of storm waves and help the mussel to turn and face into the storm.

Prime time for beachcombing is during the receding tide after a storm. Wind and waves have rearranged all objects not securely anchored to the offshore slope. The high-tide zone is littered with shells, seaweed, pebbles, and man-made debris—lumber, objects washed off boats (and sometimes the whole boat), fishermen's floats, and unfortunately an abundance of plastic containers.

While working with a camera on the beach or just looking, I often have the feeling of being watched. Looking up, I am almost certain to see a gull sailing by, relaxed and confident as it rides the up-draft that forms the cliff-top highway. The seagull, commonest of shorebirds, scavenger of the beach, is for me the most beautiful bird in flight. The unique shape of its wings, its knowledge of air currents, and its complete mastery of slow-motion flight is a joy to observe.

Gulls can walk, fly, and swim equally well. They are extremely successful in their non-specialization. They are not really oceanic birds, but coastal fishermen and moochers. They cannot glide like an albatross, fly as fast as a falcon, or turn as quickly as a hawk, but they can see colors as well as humans do and have sharper long-distance vision.

Some gulls migrate while others live the year round in the same area. At night gulls move from the beach to communal sleeping grounds. They travel up and down the coast, searching for food and shelter from storms. Particularly during winter they may be seen on inland lakes sometimes several hundred miles from the coast. Gulls seem to eat just about anything they can swallow: fish, garbage of all kinds, carrion (when nothing else is available), small mammals such as rats, moles, and rabbits. They steal eggs and the young from nests of other birds and occasionally from their own species, and they forage in meadows for worms and insects or steal wheat grains from their stalks. They carry crabs and shellfish aloft and drop them over a hard surface to break the shell. Gulls have been seen to repeat this process many times if their aim is poor and they make the drop on mud or water. While gulls are extremely buoyant in the water, they can dive for food, using their wings like fins to move them below the surface.

Although tough and adaptable, the gulls were no match for hunters who slaughtered them by the millions to use their wings as decorations on women's hats. By 1870 the herring gull of the North Atlantic coast had practically disappeared. When laws were passed to protect the gulls, the situation began to change. An almost unlimited food supply at garbage dumps and fish canneries caused a gull population explosion. As a result other nesting shorebirds, such as the tern, were in danger because the gulls eat both eggs and chicks. In 1972 the gull population stopped its increase. This may be because the fishing industry is declining and because landfill garbage disposal has decreased the food supply.

Deep in my heavy human bones I am jealous of the flying birds because I have flown only in my dreams. The true symbol of flight, the feather, has always seemed like a rare and precious gift when found along the beach. I delight in finding and photographing feathers. There is a fragile, delicate, magic quality about a feather that is a paradox, for the feather is stronger than any man-made substitute of equal weight. The veined flight feather that makes flying possible has a central shaft and two webs of flat stiff barbs. Each barb is hooked to the adjoining one by as many as five hundred barbules on either side of each barb. A single feather may have as many as a million barbules. A bird can rehook the barbules by running its beak through the feather. Down feathers provide insulation. (A hummingbird has close to a thousand feathers, while a whistling swan may have twenty-five thousand.)

Preening is a vital occupation for gulls and all other birds. When feathers are fluffed, the barbs are joined together so that the vein remains intact. A fatty substance from a gland in the tail is spread on the feathers to keep them shiny and waterproof. Large flight feathers are replaced only once a year despite constant stress on these super-light structures while in flight.

I hear the shrill call of the black oystercatcher arriving for breakfast on the reef that is now well exposed. Its long red beak is designed to pry limpets and other shellfish from their rocky homes. A long-legged willet soons joins the pair of oystercatchers to feed on the small creatures of the seaweed and tide pools. Now the brilliant midday light allows me to stalk these fast-moving shorebirds with my telephoto lens.

Waves still break with authority on offshore rocks and reefs, encouraged by a brisk onshore wind that hustles the clouds from sight over green eastern hills. There is a pervasive spirit of movement and life today that is ideal for my work with a camera. I am fascinated with moving water—streams, waterfalls, waves, rain, snow, and ice. I never tire of searching for that elusive moment when the essence of moving water is caught in two dimensions. As I work, the sky is clear except for a tiny blue-black smudge moving at the horizon. Just a little squall dulling the whitecaps in the distance I tell myself, a last remnant of the storm that would not dare to interfere with my work. But slowly a shadow of gray passes over the green and indigo shallows, and random drops of rain grow into a steady downpour. Scuttling for shelter, I duck into a shallow cave just high enough to crouch in and deep enough to protect me and my gear from the slanting drops.

The basin-sized tide pool at my feet is dancing to a steady spatter of rain. Each drop rebounds on impact, as if a whole series of minute fountains were at play. Two hermit crabs stop their bickering and head for the deep end of their pool, no doubt to get away from the unpleasant taste and feel of fresh water, for each bouncy little fountain is an added threat to the crabs and their neighbors who cannot live in fresh water. Trickles and streams descending from the rocks and cliffs above me soon join the curtain of rain. Now I feel as well as see the hard, water-smoothed cobbles that pave the floor of my cave. These green-gray and purple-gray rocks are almost identical in color and size to those embedded in the walls of this cave. The fine-textured sedimentary layers that hold the harder cobbles are tilted almost vertically. Particles of this same sediment are being carried away from the cliff by rivulets, literally right in front of my eyes, to help form another layer of sediment over similar cobbles out beyond the turbulence of the surf.

Added to the hollow report of wave against rock and the mutter, grumble, and sigh that follow are drumming and dripping rain sounds . . . *crash!* The shock of thunder jolts my perception of place and time. My awareness of brown rock, gray water, and dark sky recedes. The echoing thunder, pounding sea, and beating rain are the sounds of an earlier time when the young planet wore a cloak of clouds so thick that the sun was unable to break through to turn the new oceans from black and gray to blue.

How did it all start? There were no eyewitnesses to the birth of the earth and the oceans. We have no written record. If we are to believe that the earth was made in seven days, the days must have lasted for millions of years. There is a theory, based on evidence found in some of the first rocks that formed the solid skin of the earth and on the everyday evidence of our total dependence on energy from our star, the Sun.

This theory tells that in the beginning the Sun was the mother and natural law was the father. Gases or particles from the sun were forced together by gravity and compacted into a spinning sphere. The friction of colliding particles and the pressure of gravity on the core created the heat needed to break down existing compounds into water as vapor and the gases and minerals that we know. As this molten ball streaked through space in its new orbit, centrifugal force caused lighter materials to come to the surface and the heavier metals to drop to the center. A blanket of clouds, thick enough to obstruct all sunlight, contained the earth's water for eons while the hot surface turned rain to steam. Eternal night, lighted only by the internal fires of liquid rock, lasted until the surface cooled sufficiently for water and gravity to start working on the bare rocky slopes of those first mountains. These first rains continued for centuries, tearing at the earth, leaching chemicals from the granite, and building river arteries. Basins between the granite masses began to fill; the lightest rocks carried in the new rivers reached the ocean basins to become the first sand. When the thinning cloud cover changed the days from black to gray, there stood the oceans, ready to support the birth of the first living thing on planet Earth.

There were no eyewitnesses to proclaim this event or to explain the mystery of the parenthood of that first little organism. Could it have been bacteria from a meteorite or organic componds synthesized from atmospheric gases? The mystery remains, but fossil records suggest that this first living thing may have lived in the dark sea, a primitive bacterium that oxidized iron for food. With the return of sunlight the first plants floating in the sea used this light energy, with sea water and carbon dioxide from the air, to build what they needed to sustain life. These pioneer organisms left no trace, but their descendants, the diatoms of the sea and the grass of the land, continue to grow and to sustain all other living things.

Low Tide

After the last shower, the beach seems strangely quiet, the urgent clatter of wave on rock is hushed. Since the high tide at dawn, the water level has dropped seven feet; the sun is almost directly overhead. The sea has retreated more than eighty yards, exposing the reef, its outer edge fringed by thick-stemmed sea palms. Outpost rocks between cliff and reef wear bands of gray-black armor, thicker and wider at the impact point of breaking waves. Moving closer I see that the armor is a solid mantle of gooseneck barnacles and mussels, now tightly sealed to retain the moisture needed to survive until the tide rises. Between sand and outer reef, dark lumpy masses seem to slowly rise from the water's surface. These rocks are solidly matted by a growth of limp, slippery brown ribbons in a variety of shapes, lengths, and textures. Patches of vivid green surfgrass mottle this dark, uninviting landscape. A distant glimpse of an orange star-shaped object and something purple and spiny in one of the pools left between rocks by the receding tide is enough to overcome my reluctance to enter such foreign territory. Shoe soles patterned for rock climbing don't keep me from slipping and sliding across rocks completely hidden by damp rubbery algae.

I pull aside this seaweed covering with its iodine-like odor, heightened by the drying of the sun, to find a patina of living things clinging directly to rocks and to each other. Here are pinhead-sized acorn barnacles, limpets, snails, and many others that I cannot identify. A smooth, pink layer of red encrusting algae, like a heavy coat of paint, edges the base of the rock. Two round eyes peer up at me from a crack where a shore crab, claws extended, has wedged itself. When I turn the rock over, I expose members of the community that live in mud and gravel away from sunlight. In confusion they race for protection. Brown eel-like fish, four or five inches long, thrash violently in the soft mud and disappear; flat, gray porcelain crabs with one huge claw scuttle away. A white brittle star clings to the rock along with flat brown ribbon worms. I know that I have seen only a few of the inhabitants but I feel the guilt of an intruder and turn the rock right side up, for most of the creatures of this habitat cannot survive direct exposure to the sun.

The tidal zone is completely alive. Every niche and space is occupied by a vast array of living things that inhabit this narrow, turbulent wilderness. The variety and density of life seem to increase in the lower zone that is submerged most of the time.

This tangled mass of dark, wilted seaweed resembles a rain forest as seen from the air: it is only the top layer of a crowded community. Abundant rainfall, constant climate, and temperature with plenty of sunlight make the rain forest possible. Life exists in well-defined horizontal layers from treetop to forest floor. The population of each layer, from the butterflies and blossoms of the summit to the fungi on the shadowy humus, has accepted the limitations of a predictable environment. Similarly, life in the tidal zone is divided into layers or zones from the point reached by spray from the largest wave to the outermost reef exposed to air only by the lowest of low tides. Each of the anchored, clinging, or burrowing organisms that live here finds its niche according to its ability to withstand the shock of breaking waves, partial exposure to air and sunlight, and partial immersion in sea water. These conditions for life seem too severe to be practical and yet there are more varied life forms in the tidal community than in the rain forest. Here at the edge of the sea the resilience, determination, and adaptability of the living organisms are vividly revealed in the designs and systems developed for survival.

With few exceptions, the plants and animals here on the beach are sea forms that have learned to tolerate part-time exposure to air. This is possibly a key to the great success of tideline life. The ocean contains more varied and numerous life forms than does the land. This might be expected, since life in the sea exists from the shallowest edges to the deep ocean bottom, while on land life is limited to a narrow band between the roots and tops of trees.

Sea creatures have another great advantage: it is easier for them than for creatures of land to maintain the essential salt and water balance of their body fluids because sea water contains the required minerals and liquid.

Awareness of the inevitable flood that will follow an ebb tide brings a feeling of urgency to my exploration of this narrow wilderness, only a few feet wide but thousands of miles long. Near the rock I have just turned over, my shadow crosses a shallow pool, breaking the sky's reflection on its surface to reveal the bottom covered by a solid mosaic of green, purple, and orange shapes. Reaching down through the layer of clear water, I feel the rough, leathery body of a five-rayed starfish. As the water around my hand stills, I see the suction-cup feet of the starfish hungrily embrace a mussel shell. Close by in the same pool lives a flower-like green sea anemone. Its delicate petal-shaped tentacles contract around my probing hand. Although I feel only a tingle from this sensuous grasp, a small fish or crab would be caught by its stinging fingers. Mindful of a series of sharp spines, I try but fail to lift a bright purple sea urchin from the pool. It is roundish, about the size of my fist, and firmly established in a hole that just fits its shape. Long, narrow probes, shaped something like the starfish's suction foot, shoot out from the body toward my fingers. Soft yellow and red sponges complete the pattern, which is grouted with pink, encrusting algae.

Because of intense competition for living room on the rocky beach, open space in this pool is limited to the sand-covered floor at the seaward edge, where the raised prow of the reef breaks the impact of scouring surf. Shifting particles of sand make a poor foundation for a permanent home.

Shells of the purple turban are scattered over this sandy arena. Each snail lives and moves with the security of its homemade castle. An armored door, called the operculum, carried on its single foot, is moved into place to seal the entrance if hunters threaten. This waterproof barrier also protects the snail from sun and air as the tide ebbs.

Suddenly a snail shell moving at several times the normal snail's pace catches my attention. As it stops abruptly a spidery, crablike creature pops from the shell, bursts open, then steps right out of its skin like a diver struggling from his wet suit. No chance for a closer look as the creature ducks back into the shell, blocking the entrance with two claws streaked with red. A hermit crab has molted before my eyes! This glimpse revealed two crab claws, one larger than the other, eyes on the tips of movable stalks, four blue-and-white-striped legs instead of eight, as on other crabs, and a dark, rubbery, crooked tail.

I note that most of the snail shells in sight are occupied by these little crabs. They are everywhere once I start looking. Undoubtedly they appear to be so numerous because they don't have to hide and because they lead such busy lives, eating, fighting, mating, and searching for bigger homes as they outgrow their shells.

Like all crabs, the hermit has a hard, segmented outer covering called an exoskeleton (in contrast to a fish or mammal, which is supported on the inside by a backbone with attached ribs and interior supports for the extremities). This exoskeleton not only supports the organism but offers protection against a host of meat eaters. Most crabs have a blunt, well-armored rear. The exception is Pagurus, the hermit crab, who is of the group Anomura, meaning "with an unusual tail." For good reason, Pagurus is very sensitive about his behind, which is long, slender, and unprotected by the armor that covers the rest of him. Snail shells seem to be just the right size, shape, and weight to serve as comfortable mobile homes for these little soft-tailed crabs. Their tails have a twist to the right that matches the internal contours of the snail shell, while their claws fit neatly into the entry hall to block intruders. I find myself wondering if their body style has adapted to life in a shell, so that they no longer need eight legs and built-in rear protection, or if they have survived in such numbers only because snail shells are so available.

Pagurus has a varied appetite. During his larval stage, spent floating near the surface with other small plants and animals, he feeds on single-celled plants until he is large enough to eat copepods and other small animals. As an adult he helps keep the tide pools sparkling clean by eating any plant or animal particles, alive or dead, that his strong right claw can grasp. This scissor, plucker, pincer claw is forty times as strong as a human hand, in proportion to its weight and size.

The outer or exoskeleton is a most successful body design for life on a planet full of hungry predators. The crabs are just one branch of the phylum called Arthropoda, which includes the insects and spiders on land, and the shrimps, barnacles, lobsters, and crabs in the sea. Over four hundred thousand species of these arthropods account for eighty per cent of all the animals of our world. Because the hard outer covering of the arthropod does not grow, it is shed periodically to make way for the next larger size. The skeleton is formed as a secretion from underlying membrane; a new outer covering is formed while the old one splits and loosens. Rapid growth takes place before the new covering completely hardens.

The hermit crab that just molted is probably beginning to feel the housing pinch and will soon be on the lookout for a larger model. I will try to imagine being with Pagurus on his quest for a new home.

The coming flood tide will disturb the glassy blue roof of the crab's tide-pool world and stir drooping algae forests into swirling motion before Pagurus leaves the sandy clearing. Surging particles of foam cast fleeting shadows across the tide pool, followed by rippled patterns of sun, beamed through moving water. Through wide-field eyes, the little hermit crab watches bubble clusters skim the surface in the wake of breaking surf. Because water carries the vibrations of movement more perceptibly than air, the hermit crab's white-banded antennae can sense that his relative the striped shore crab has left his rock-crevice refuge to amble upward with a sideways gait. He watches this crab with the green body, striped with narrow purple lines, disappear through the water-air boundary into the realm of air, wind, and flying spray. The hungry shore crab is off to find his share of the floating food served up on the reef by the flow of water moving in from the open sea. This crab can stay out of water for quite a long time because his gills have adapted to his choice of hunting grounds.

On the floor of the pool, Pagurus watches snails and other hermit crabs move toward the cover of surfgrass. He sees the limpet colony stirring on their home rock, as they prepare to graze over algae-covered rocks. He feels the rhythmic turbulence overhead and is refreshed by the cold aerated water that it brings. A cresting wave breaks over the little crab's pool, casting a curtain of gleaming bubbles before his eyes while stirring sand and sediment. These bubble masses bring precious oxygen but they decrease visibility during the tidal transition.

The incoming tide is also a time of danger. Small tide-pool creatures have learned that a passing shadow could be sculpin (the large-mouthed rock fish), a large cancer crab, or one of the surf fish, all moving in with the tide to prey on tide-pool citizens. The busy work of survival, geared to the tidal rhythm, has started at all levels in the little pool. The almost invisible ghost shrimp darts along the surface feeding on single-celled plants. Under rocks flat worms, tube worms, eel-like blennies, crabs, and brittle stars are busy. The filter feeders, barnacles and mussels, extend feathery strainers. The ocher starfish again mount their inevitable attack on the outskirts of the mussel village. Slowed by the effort of changing his clothes, our hermit crab lingers: he must find a larger home and he is hungry but must move with extra caution while his new shell hardens. The decision is made for him by a small white-and-brown-striped convict fish that shoots past him, one length ahead of sculpin's open jaws. The crab scoots sideways into the surfgrass and the fish loses a meal as his teeth barely graze the shell. Safe for the moment but still shaken by the attack, Pagurus pushes on into the heart of the seaweed forest. There is shelter from swimming hunters here at the shadowy base of brown, green, and red stems and fronds that reach for the surface far above. Held snug against the surging tide by the knobby arms of a holdfast and protected by retracted pincers that neatly block the entrance to his home, the exhausted hermit crab rests. Pagurus could move swiftly to escape the fish because the shell he must carry and his protective outer skeleton weigh little more than water.

As the restless tide dashes over the tide pool, energy from the afternoon sun bursts through the surface in quick flashes to highlight glistening swirls of seaweed. Streams of passing bubbles reflect patches of yellow and red sponges. This light show, directed by turbulence and sunlight in the shallow pool, finds the community at the peak of its activity cycle. Stomachs must be filled, homes found, tunnels enlarged, lovemaking finished, the essentials of living taken care of before the tide changes again.

Hunger pangs soon awaken the little crab. For Pagurus the gracefully swaying forest above, gleaming with refracted light, means that food will be plentiful. The tide-pool soup pot is being stirred from the bottom, and with his large appetite the crab will not be choosy. Pagurus starts eating bits of red algae that drift by his sheltering holdfast, then moves toward more of his favorite vegetables growing nearby. As he pulls with his claw, the plant and the whole garden vibrate and turn completely until he is staring into the beady eyes of the decorator crab, who is large enough to swallow him in one bite. With one eye on the decorator crab's great pincer claw, Pagurus backs away swiftly. Only a sleepy, hungry crab is easily fooled by the traveling garden that masks the decorator crab. With a special glue secreted in her mouth, this crab attaches growing plants to her body. She pulls the whole plant with its holdfast from its home so that it will continue to grow. This array must be replanted each time the decorator crab molts and grows a new shell.

As Pagurus moves at top speed from the huffy, masked crab, he is delighted to find that a feast is waiting at the edge of the seaweed forest. Sculpin has caught a cancer crab crossing the open sand. Because fish are not tidy eaters there is plenty left for hermit crabs, flatworms, brittle stars, and the rest of the tide-pool cleanup crew. Hermit crabs have short tempers, particularly when hungry, so Pagurus will have to fight with his brothers and sisters for his share.

Between eating and using his claws to box with friends, which all hermit crabs appear to enjoy, there is not time to find a new home before the next slack low-tide period. And so the rhythms of ebb and flood, the light and dark cycles of day and night pass, until finally, like hiking in shoes a size too small, the discomfort of living in cramped quarters becomes unbearable for growing Pagurus. All the shells in his home territory are worn and broken, or too small, or already occupied, so he must journey the length of the forest, then skirt around mussel-covered rocks to a distant place where the giant green anemones live.

If he had not been backpacking a pinched tail, the hermit crab might be enjoying the many different shapes, colors, and sizes of growing things in the tide-pool forest: the saturated green of hairlike eelgrass, the heavy brown stems of oarweed flattened into amber blades near the surface, paper-thin red algae pulsing with every movement of surrounding water, and an occasional glimpse of round and cone-shaped floats bobbing on the surface to support each stem.

Moving along with an eye out for shells, the crab passes a delicate plant of orange filigree without even noticing. This is the ostrich-plume hydroid, an eating tree made of plant-like colonies of animals living together in graceful symmetry. One group of flower-like individuals captures food with stinging tentacles, while another class of vase-shaped organisms takes care of reproduction for the colony, producing young called medusae. These free-swimming larvae wander on the sea's surface before settling down to form new hydroid colonies.

All through the forest, holdfasts, stems, and rocks are ornamented with small structures patterned after the finest lace. These are the Bryozoa, colonies of animals each with its own little protective chamber. Every member of the colony prepares its own meals by straining the tide-pool soup with minute tentacles.

The quest for a new home is a long, tiresome journey for the little crab. Before he reaches the end of the forest, Pagurus's big, round eyes are on the lookout not only for shells but for food. He sees but ignores the soft, succulent snail without a shell that grazes through the algae unafraid and apparently unprotected. Daubed with fluorescent orange and purple pigments, the nudibranch is the slow-moving "butterfly" of this enchanted forest. The brilliant coloring of this sea snail carries a message to all predators that the wearer of these bright robes tastes very bad. The nudibranch has a very large relative in the forest that is called the sea hare because it has feelers that resemble rabbit ears. This large green-red-brown sea slug can be fifteen inches long and weigh sixteen pounds. It has an internal shell, a thin plate in the mantle. The sea hare's coloring blends well with the tide-pool forest, where it grazes on red algae, but if attacked, the sea slug ejects a purple fluid that conceals its location while repelling the attacker.

Pagurus emerges from the forest into a narrow underwater valley which protects him and more permanent residents from the churning surf of the outermost tidal zone. Unknown enemies may lurk in these shadowy depths, but at least the rocky walls of the valley offer protection for visitors, as well as living space for the robust community that inhabits this slanting gap on the edge of the reef. No room for plant holdfasts on these rocks, as a tight mosaic of clinging animals has won the competition for space. Clambering over and around masses of barnacles and mussels is slow work for the little crab. The homemade mooring lines that secure the mussels make a difficut obstacle course, and starfish with their hairy pincers must be avoided.

Traversing patches of bright sponge, Pagurus feels the current generated by these living filters. Between an interlocking network of skeletal fibers, individual cells of the sponge, each with a moving whisk, move water through the body, where small particles of nourishing sea soup are trapped. A sponge only four times the size of the hermit crab's shell uses over two million pumping cells to filter more than twenty quarts of sea soup in one day.

With giant green anemone finally in view, the crab's antennae pick up vibrations of danger above. Without focusing his eyes on the dark shape diving from the surface, Pagurus ducks under a ledge and wedges himself between a rock and a giant abalone shell. Trailing an arc of bubbles, a sleek sea otter drops lightly toward the valley floor looking for a purple or red urchin, an abalone, mussel, or any other shellfish that his nimble paws can loosen. A lifetime of diving has taught him where to look, but the abalone, alerted by the hermit crab, has clamped its muscle against the rock and is safe from the probing otter. Not to be denied his favorite meal, the sea otter picks out a flat rock, pointed at one end, and with this tool pries an urchin from its crevice and then ascends with both rock and urchin. The sharp tapping of shell on rock resounds through the valley as the sea otter, floating on his back, breaks the shell against the rock and then eats the urchin.

Meanwhile, hiding behind the red abalone, Pagurus feels water jetting from a series of holes in the abalone shell. These holes allow water to reach its gills and so the holes perform the same function as our nostrils. The abalone, largest relative of the limpet, has survived here only because it is well hidden from its most destructive predator, the human hunter who often kills for sport rather than survival.

Yet another very distinctive member of the mollusk group lives under this same ledge because it is rather sensitive to bright light. The chiton, or sea cradle, formed of eight butterfly-shaped plates, is held together around the outer edge by a strong, flexible girdle. Like the limpet and the abalone, the chiton has one large creeping foot that carries it over the rocks to feed on algae. The lovely oval chiton shell can be black, brownish, greenish, orange, or red. When the chiton shell is washed up on the beach it is curled into the shape of a cradle.

At last free from immediate danger, Pagurus finds a cache of shells at the deep end of the pool, among anemones. He locates a small cavern where shells have become lodged when storm currents swept the valley floor clean. There are enough sizes and shapes here to outfit a whole colony of hermit crabs. With practiced eye he searches through the used shells, measuring with claws, appraising shape, age, and color. Each likely possibility is turned and examined inside and out. The caution of survival precludes him from trying on any shell but his final choice. Finally a gleaming purple turban is selected, handsome in every way and certainly worth the long trip; but suddenly his new shell is in the claws of another crab who has arrived unnoticed but with the same intent.

With his first furious charge, Pagurus sends his rival sprawling, but before he can slip out of his old shell and into the home of his choice, he is attacked with swinging claws. Shells fly in all directions as the battle rages. Seeking an advantage, the newcomer jumps from the cavern to grapple from above, but lands on the outstretched tentacles of an anemone. Even before he knows his error the stinging arms of this beautiful plantlike animal have begun to envelop the crab, until he disappears through the slit-mouth. And thus Pagurus achieves his bright new home.

The name sea anemone does not describe the nature and habits of this animal with petal-shaped tentacles. The only thing flower-like about the anemone is its radial symmetry and its beauty. The hollow tentacles are lined with stinging cells that inject poison-filled needles into small fish or large plankton. The outer surface of the tentacles is covered with almost invisible hairlike cilia that constantly beat in the direction of the tip of the tentacle. When something edible comes along, the tentacles contract over it so they point toward the mouth. The current created by the beating cilia pushes the food in that direction. When the tentacles are spread, the cilia-created current keeps them clean and free from obstructions that might interfere with the stinging cells. Cilia lining the gullet carry food into the stomach. Except when the anemone is eating, these cilia move to maintain a flow of water into and out of the interior cavity, so that a steady supply of oxygen comes in and carbon dioxide and other wastes go out.

The body of this animal, shaped like an up-ended tube, rests on a muscular foot that can cling so tightly that the flexible body will be torn away before the foot pulls loose. A central mouth closes the upper end of the tube body. The interior cavity is segmented into compartments for digestion and circulation. When under attack or exposed to sun and air by changing tides, the anemone retracts its mouth into the central cavity, and a fold of skin makes a collar to hide the tentacles, so the whole body is tightly closed.

Anemones flourish on beaches of all the world's oceans in a great variety of size and color. The green anemone grows to be more than six inches in diameter and can live for hundreds of years. Algae living within the tissues of the anemone contribute the handsome green color to the host. The anemone supports the algae and feeds on nutrients synthesized by the algae. This is a successful symbiotic relationship—both partners benefit.

As I lift my hand from the pool a shower of clear drops flashes in sunlight for a moment before arching back to the surface. The drop left on my lips is more than salty, it's bitter with a taste as hard to describe as its smell. The smell of seawater is that of a delicate soup with many seasonings; not appetizing to the human palate or nourishing either—except for the larger morsels, fish and shellfish—but a soup that feeds most of the living things of our planet. Starting with a clear broth containing at least traces of all the natural elements, this soup is enriched with plants and animals that are food for each other. The drop of water that I just tasted probably contains thousands of single-celled plants. On this coast, the nutritious sea soup is ladled over a rocky shore that provides a firm foothold for creatures needing a permanent home. This rare combination of room and board attracts and supports the thriving tideline community.

Looking again into this brilliantly clear pool, I remind myself that a host of invisible plants and animals float here. A glance toward shore confirms that all the space on each rock facing the sea is encrusted with living things whose needs are met by wave-borne food. Remembering pictures taken through powerful microscopes, I try to visualize what I would see in a drop of surface water being rushed by the surf toward barnacle-covered rocks. If my vision could perceive only this drop of water, I would be dazzled by an array of intricately detailed, mostly symmetrical shapes, all unfamiliar except for a few which look rather like shrimps. The largest organism, a copepod about the size of a pinhead, has oar-shaped legs and hairy feelers. The smallest, a diatom less than one thousandth of an inch long, is a tiny, jeweled pillbox. This "micro-sea" is hurtling toward a giant net that strains the water as it goes by, but most of the swimmers in this drop are far too small to be caught by the barnacle that has extended its six pairs of fringed sweepers.

The organisms in my imagined drop of water are called plankton, meaning "wanderers." All marine creatures that are too small or weak to swim, anchor, or burrow drift with the tides and currents. Most of the plants and animals of the ocean are drifters in the sea at one time in their lives. Because the plankton are not evenly distributed through the sea, it is difficult to estimate their number. However, the average number of plankton counted yearly for fourteen years in the same Atlantic bay was over twenty thousand per cubic foot of water. When separated by a centrifuge rather than a fine-meshed plankton net, at least twelve and one half million single-celled plants were found in each cubic foot of water.

The single-celled plants called diatoms are the smallest and most numerous of the plankton. Ten thousand different species of diatoms amount to more that half of the billions of drifters in the sea. They surpass in edible volume all other land or sea plants. We drink billions of them, as they live in fresh water also and are not poisonous. With the grasses of land, diatoms are the largest and most important industry on the planet because they are at the beginning of the long succession of living things that eat each other.

The outer wall of this one-celled plant is silica; shaped like a box with a tight-fitting cover, it contains a tiny drop of oil with a fishy smell. Silica, a basic ingredient of rock, is drained into the sea in great quantities and used almost exclusively by diatoms. These jewel-like structures, designed to withstand great water pressure, are symmetrical and as diverse in shape as snowflakes. They are often cylindrical or have three, five, or six sides. Outer walls are concave, convex, corrugated, or double-thick for extra strength. Diatoms can be needle-shaped, lens-shaped, or eye-shaped.

After a life-span of only a few days during which it reproduces by division, the microscopic silica structure will slowly sink, to feed plankton at lower levels or to reach bottom and become part of the diatomaceous ooze. Water pressure and the passage of time will transform this ooze into a chalky rock that is used in industry as a polishing material or a filter. The diatom may completely dissolve before reaching bottom if the sea depth is a mile or greater.

However, most of the living diatoms will be eaten by animal plankton, very likely a copepod. The five thousand species of copepods make up the largest group of multicellular animals on our planet. These animals in all their variations are larger than diatoms. Generally shrimp-shaped but with long feelers or antennules, copepods have a great advantage over the diatoms: they are self-propelled in a jerky way and have filtering appendages that vibrate from six hundred to two thousand four hundred times a minute, producing a current that brings them food.

Diatoms are the producers of energy for all animals in the sea. Copepods are in the first rank of consumers. A well-fed copepod may contain over one hundred thousand diatoms, while the anchovy that feeds on the copepods might consume from five thousand to eight thousand copepods to feel satisfied. A humpback whale needs at least five thousand anchovies to satisfy its hunger pangs. Thus the meal for a medium-sized whale indirectly requires a few million diatoms. However, this web of consumers has its exceptions. The great blue whale, much larger than the humpback, consumes only plankton, which it strains from the sea as it swims. Diatoms, copepods, anchovies, and whales are parts of a vast food web that includes all the animals of the sea. This web is shaped like a pyramid with far greater numbers of the smaller animals at the base, while at the top of the pyramid are the tuna, the sharks, and the whales, relatively small in number.

The task of the invisible diatom is great, beyond our comprehension. In addition to maintaining its own population, it supports all the herbivores at all levels of the sea, which in turn support the carnivores. Diatoms use the sun's energy to accomplish their mission. The struggle for a share of the energy from the sun is the essential reality of life on planet Earth.

Sunlight is converted into food by molecule-sized laboratories called chloroplasts that are part of the structure of all plants. A combination of water, carbon dioxide, and light is transformed into starches, sugars, and oxygen. The process is instantaneous and so complex that it seems magical. Molecules of water and carbon dioxide break down into hydrogen, carbon, and oxygen, which react with each other to form a series of compounds which finally produce free oxygen and basic plant foods that are carbon compounds. This process called photosynthesis is carried one step farther by diatoms. The final products of their reaction are fatty acids and waxes. Under ideal conditions, diatoms reproduce and grow at such a tremendous rate that many square miles of the sea may appear to be covered with oil.

Oxygen is just as necessary for marine animals as it is for animals who breathe air. Diatoms satisfy this need as well as provide the basic food supply. The green-plant pigment chlorophyll is the link between the sun and all life on this planet. Each diatom contains one or more molecules of chlorophyll. Nutritional needs other than sunlight are met by sea water; the minerals needed for growth—silica, nitrogen, and phosphate—are here in solution. The other essential, carbon dioxide, is found in sea water in a concentration one hundred times greater than in an equal volume of atmosphere.

Life in this rich environment has its problems too. Some of the minerals needed by diatoms are only available in very small amounts near the surface, in quantities much less than in good soil. Availability of sunlight for photosynthesis is limited in the sea, for while most of the radiation coming to earth falls on the sea, much of it is reflected from the surface or converted to heat. Many wavelengths of light are not usable, and the intensity of light decreases rapidly as depth increases. The microscopic size of diatoms helps them absorb sunlight and minerals efficiently, because the ratio of surface to volume increases as the body becomes smaller. Diatoms' greatly extended outer surfaces relative to body size also help to keep them floating near the sunlit surface. In summer when the water is warm and density low, the shells of diatoms sometimes extend their exposed surfaces to add buoyancy. Diatoms are able to fulfill their destiny by making the most efficient use of only two hundredths per cent of sunlight reaching the sea.

In the time remaining before the flood tide mounts its first surge of fresh, nutrient-laden water over the lip of the reef, each member of the exposed tideline community must tolerate the ordeal of air and sun in its own way. The anemone has contracted its flexible body, leaving only a small area exposed to the sun. Starfish and the shore crabs have retreated behind a shield of thick wet seaweed. In shallow pools directly exposed to sunlight, the temperature has risen, the salt content has increased with evaporation until little oxygen remains in the still water. The decaying odor of dry seaweed hangs over the reef. A temporary truce between the hunted and the hunters is in effect as vital processes slow down to wait for the change that always comes.

Now is the crucial time for close observation. I look forward to fresh discoveries, remembering strange new treasures that I found in tide pools as a child. I have observed my own children experience the excitement and joy of discovery. Perhaps the most rewarding view of any wilderness is that of childlike vision. The hunter in his wet suit sees easy access across the reef to submerged ledges and crevices where he hopes to find abalone overlooked by a horde of hunters before him. The collector sees a seldom-exposed rocky plateau that may be the home for living things that can be preserved in alcohol for closer examination in the laboratory or just tossed into a bucket to take home because "they are pretty." Near where I live, intertidal reefs have been completely stripped of life because they are accessible to marine biology and zoology classes and to hunters. Hunting with a camera means having a two-dimensional memory. Images, feelings, and the appearance of reality can be shared with a minimum of intrusion into the environment.

Against a background of waves and open sea a large pool, its floor sloping down to the very edge of the reef, is the final barrier to my wandering. A sudden flash of intense orange lights the shadow of an overhanging rock at the deep end of the pool and is gone. Something has disturbed the quiet, clear water, and the shapes of ocher star, urchin, and eel grass appear as wavering bands of purple, yellow, and green. Another shape is there, large, round, and bright in the center. Could it be?—Yes! It is the great sunflower star, the largest and most beautiful of all starfish. There is no time for a closer look, as it moves again. I scramble for a longer lens and faster color film to photograph my find. The picture will show nineteen arms radiating from a large central disk outlined in fluorescent red-orange, as if the disk radiated light. Over two feet in diameter, the body seems to glide with no effort over the uneven floor of the pool. I am elated with this discovery, which is of great beauty. This day is now very special. One new find or one truly important picture is enough for any day. Sunflower starfish can have as many as twenty-four rays and grow as large as four feet in diameter. It has a very soft skin and a body that collapses immediately when removed from the water.

The starfish is an incredibly successful example of adaptation to the rigors of tideline life. It and its relative the urchin have a body design and life support system that is different from all other animals. It has no eyes, brain, spinal cord, or blood; yet it has a most efficient built-in hydraulic system which powers thousands of tube feet. Instead of taking food into its body, it reaches out with its stomach to digest its victims.

Here in the low-tide zone, several dozen starfish are visible to me right now. Their radial body plan and bright colors—orange, purple, and red—are hard to ignore. They have almost no predators to hide from except the bright sun, although I once saw a gull repeatedly fly with a starfish in its beak and drop it onto the rocks to make it more palatable.

The skin of the nearby ocher star has a hard, rough feel. Its skeleton is a loose network of bony plates and rods that allow flexibility and also provide protection. Projecting from this meshwork is a series of movable spines used for defense. Between the spines are many small pincer-shaped structures, the pedicellariae, that open and close like pliers and scissors. Short stalks that may be extended and moved attach the pedicellariae to the body. The multipurpose little tools are used for cleaning the body, for sensing, for protection, and for seizing prey. When I put the back of my hand next to the starfish, many of these tiny pincers grab my hair, so that I can lift its body by raising my hand. A small crab with many bristles could be trapped easily in the same way.

When a starfish moves, it reaches out with its many tube feet and pulls its body along. Each of the thousands of hollow feet is tipped with a suction disk and is moved by the hydraulic action of an individual, muscle-compressed water bulb. Each bulb has its own valve which controls the flow of water. At the same time each valve is connected to a radial canal running the length of the arm, which is filled from a circular canal in the central disk of the body. Filtered sea water enters this system through an opening in the central disk. The circular ring canal is lined with tiny oarlike hairs called cilia that keep pumping to maintain the pressure needed to make the system work. It is almost impossible to pull a determined starfish from the rocks without using enough force to break off some of the tube feet.

These water-powered feet not only guarantee security for the starfish in the wildest surf, but are also the weapons that make it the fiercest predator of the tide-pool community. Although the starfish prefers the thick-shelled mussel, it will eat acorn barnacles, limpets, and snails. The hungry predator humps over its victim so that the tube feet can pull at right angles to the shell until inevitably the exhausted mussel relaxes and the starfish is able to push its stomach into the shell to enjoy a meal. As soon as the valves of the shell are slightly open, enzymes are injected so that the meal is tenderized by the time the stomach gets there. While the little water-filled tube feet of the starfish are not nearly as strong as the muscles holding the two valves of the shell tightly closed, there are hundreds of them, backed up by the pulling power of several radial arms that do not tire.

The starfish breathes through tiny gills that are membrane-covered projections in spaces between the meshwork of the skeleton. Carbon dioxide, collected by a liquid that fills the body cavity, is exchanged for oxygen from sea water through this thin membrane. Tiny cilia move to keep circulation moving on both sides of the membrane.

Egg and sperm cells are extruded by the female and the male from pores in the arms, to be fertilized in the sea. Baby starfish, in their larval stage, float with the plankton. Certain species of starfish spawn under tideline rocks. The female collects a cluster of fertilized eggs which she holds with her tube feet until the larvae escape from the thin sack in which they are enclosed for the first three weeks of their life. For forty days more she protects them with her body, holding to rocks with only the tips of her arms. She does not eat during the entire period that she stays with her young.

Starfish, with their simple, efficient design for living, have successfully colonized all the oceans of the world. The balance of life in some parts of the sea is tipped precariously in their direction. An army of these predators is now eating its way across the Great Barrier Reef near Australia, threatening one of the largest underwater communities in the world.

One of the many variations in the widespread starfish family is the brittle star, about three inches in diameter. With its slender arms and a perfectly round central disk, the brittle star lives only under rocks and in crevices. It is sometimes called the serpent star because the arms are flexible enough to writhe like a snake; however, the arms break off very easily if attacked. Since the tube feet of the brittle star are too small to help it move, the flexible arms are used like oars to propel the body.

Looking out to sea from the edge of the reef, I see the swirling patterns of a brown broad-leafed plant, a Laminaria, supported on the surface by its own built-in circular floats. I see only the tops of these structures, which grow up from a bottom as deep as sixty feet. Largest of all sea plants, this giant kelp grows on the edge of many of the earth's oceans. As I look down at the massive floating kelp, I am reminded of once flying over a redwood forest and seeing only the treetops, each isolated by heavy white ground fog.

This plant, which can grow two feet a day to reach a length of three hundred feet, is supported at its base by an anchor called a holdfast. The holdfast can be eight to ten feet thick and four to five feet high. Its only function is to hold the plant firmly to the bottom. Sea plants have no need for roots, for, like the tiny diatom, they take moisture and nutrients from the sea around them. Consequently, kelp need not spend energy building a thick, rigid trunk or stem for support; water, hundreds of times denser than air, supports the structure with the help of hollow floats. To flourish in the sea, kelp or any other seaweed must have a flexible but very strong stem, or stipe as it is called, to withstand constant movement at the surface. Because sunlight, the essential element of growth, is most available near the surface, the stipe carries the nourishing products of photosynthesis downward to support growth near the dimly lit ocean floor. To do this, the innermost core of the stipe carries a pipeline composed of long cells with end walls that have many small holes. An outer core of branched and intertwined filaments supports this pipe. Cells on the surface of the stipe continue to divide and grow as long as the plant is alive.

Kelp produces a rare chemical called algin that gives the plant structure great strength and resilience. The leaves or blades live for several months before being replaced by new growth in a continuous process.

Most sea plants, including kelp, are classified as algae; they have no true roots, leaves, flowers, or seeds. I have looked for a more specific definition for alga but have yet to find it. Alga is a simple primitive plant form. Certainly plants of the sea are very successful in their simplicity, for ninety per cent of the world's oxygen is produced by marine algae, from thirty thousand different species ranging in size from diatoms to kelp. The natural green color of chlorophyll is masked by brown pigments in kelp as an aid to photosynthesis in the underwater spectrum of sunlight.

Some kelps are perennials while others grow to maturity in one season. Sometimes, during the spring, millions of spores a minute begin to develop on blades near the base of the plant. If the bottom surface and light conditions are adequate, a few of these spores attach and grow from the sea floor.

A forest of giant kelp offers refuge to a host of other plants, fish, and sea animals. Schools of fish-fry play the serious game of hide and eat with larger fish in its shadowy protection. The kelp holdfast shelters a large community of its own. Like forests of the land, this is a self-contained environment.

By midsummer the kelp forest has grown to become a natural breakwater along some coasts. The force of waves is greatly diminished, and exposed rocky shores are calm for weeks at a time. After these long days of sunshine and growth, the first powerful winter storm will tear and batter the full-grown kelp beds, leaving beaches piled high with decaying remnants.

In parts of the world where kelp is harvested for its algin and other useful components there has been dismay in recent years because of the decreasing supply, even in areas untouched by man. Recently marine biologists discovered that the well-being of kelp, of spiny sea urchins, and of sea otters is closely related. One of the favorite foods of sea urchins is young tender kelp plants just starting to grow up from the sea floor. Urchins were observed eating their way through the kelp forest at a rate of thirty feet a month, leaving only bare rock behind. At the same time sea urchins have always been a staple in the diet of sea otters. When the sea-otter population of the world was almost completely destroyed by hunters, sea urchins could feast on kelp undisturbed by their traditional predator. Fortunately, with protection by law, sea otters have prospered and are rapidly increasing, nourished, no doubt, by a large supply of urchins. The kelp forests, too, are starting to expand and return to their former grandeur. If man can manage to stay out of the way, the balance of this three-way interaction will perhaps be reached again.

I have never thought of the sea urchin as an enemy but rather as a rich, finely wrought treasure, to be enjoyed for its strange beauty. The urchin, with its deep purple color and spheroid shape, covered with a symmetrical pattern of delicate spines, bears no resemblance to its close relative the starfish.

Sea urchins live in all the oceans and have varied coloring and size. The most common urchin is purple and from two to four inches in diameter. It retains the circular body design of the starfish; but the skeleton or test, as biologists call it, is rigid, as its component plates are fused together. Each of the movable spines, used for defense, has a concave base that fits over rounded knobs on the test. These spines continue to grow as they are worn away. Like the starfish, the urchin has a series of water-powered tube feet. However, the urchin's feet are also used as breathing organs. Two tiny canals extending through holes in the test connect each tube foot to its water supply. The handy tools called pedicellariae of the starfish are also used by the urchin. Each of these tools has three small clamping jaws that can be extended on thin stalks or retracted. The array of pedicellariae is used for defense, for capturing small prey, for cleaning the body surfaces of debris, and for holding bits of shell as ballast in rough surf.

The urchin moves about with surprising ease, pushing and pulling with spines and tube feet. On exposed, wave-swept, rocky coasts, urchins excavate holes in the hard rocks that exactly fit their bodies. The reef where I stand is honeycombed with little urchin-sized caves. It is somewhat of a mystery that they are able to dig in rock that is harder than their spines. Some marine biologists maintain that they keep moving around, scraping with spines and jaws. Another theory suggests that they secrete a chemical that softens the rock.

The urchin eats with five chisel-shaped jaws that are part of a complex set of interlocking plates forming a powerful chewing system. This curious jaw structure was discovered and studied by Aristotle and so is called Aristotle's lantern, because of its shape. Besides kelp, urchins will eat just about any plant or animal material, living or dead. Because space is at a premium in the tideline world, the urchin is host to other much smaller animals that live within its digestive system. As many as forty worms have been found within one urchin. This relationship is know as commensalism—one organism benefits while the other member is not harmed.

Again the quiet surface of the pool beside me is disturbed. Only a narrow raft of foam breaking up into glossy bubbles mars the surface. The gentle cluck and gurgle of slowly moving water behind me would go unnoticed if it were not a new sound added to the rhythm of wave on rock below the reef. Sound has receded with the tide, leaving a calm broken only by the harsh call of the black oystercatcher who feasts on limpets pried from the rocks with its crowbar of a bill. Turning again to the pool, I see that the new sound was the first surge of the flood tide; fingers of foam moving across the pool to erase the sky image reflected on the surface.

The sun is now west of the overhead midday point. This place on the moving earth has passed through the trough between two great planetary waves that form the tides.

On very wide, flat, protected beaches, I have watched the incoming tide advance steadily, but on this uneven rocky reef the slow, inexorable flooding is quite imperceptible. In the constant ebb and flow, a surge will gain ground and then lose most of it. The flood rises quietly in channels among the rocks, surrounding the high places before capturing them completely. There will be no hurrying, for the tide moves with the stately rhythm of the rotating earth.

Flood Tide

A new flood has breached the barnacle-encrusted rampart of the outer reef. Sea-palm blades toss and swirl in renewed turbulence. Foam obscures the image of textured purple urchins that I see through the camera lens. Turning toward shore, I glimpse the trail leading through meadow and forest to soft green coastal hills. The cliffs, outlined by wind-stunted cypress trees, seem far away. Only seven hours ago as I stood on this cliff, a wave raised its foaming crest high above me. The starfish, urchins, and hermit crabs in the pool close by were under six feet of rushing water. The flow of tide, day, and weather has profoundly changed the look and feel of this place.

I most enjoy the effects of this constant change when exploring the tideline with a camera. Infinite variations of light, movement, color, and texture make photography here exciting, although often frustrating. Intense observation increases my awareness of moment-to-moment change, for the same combination of visual elements will never occur again in just the same way. Dull-brown seawood becomes iridescent as it reflects thinning fog; the rich color and sparkle of pebbles fade as foam recedes; rocks disappear under a film of rising water.

From reef to eastern horizon, plants carpet the land except for the strip of beach sand. Even the cliff face is colonized by sea fig and cliff lettuce, plants that store water against drying winds. Above the cliff, meadow grasses ripple in the breeze; red paintbrush, orange poppies, and blue lupine are evidence of the flowering and seed-making cycle. Here on the reef muted brown, green, and red plants of the tideline meadow are abundant but bedraggled. These seaweeds are the surviving pioneers of the plant world that moved from sea to land. Land plants need strong, flexible supporting stems, while seaweeds lie limp and prostrate when exposed to air. As the rising tide lifts their blades, they move gracefully, with the whim of water, back into their natural element.

Like their tiny relative the diatom, all the seaweeds here except the long-haired surfgrass are algae, adapted to grow in the sea and survive exposure to air during the ebb. The ocean environment provides the carbon dioxide needed for photosynthesis and the phosphates and nitrates for growth. Fewer than a hundred species of flowering plants live in the sea, in contrast to over two hundred thousand species on land. Fossil records indicate that most of the characteristics of land plants developed from seaweed algae. Filaments were first welded to form ribbon tissue. Tissue was layered to form body and substance. Later a more complex structure containing the first roots, leaves, and stems evolved.

The upper tidal zone of this beach is the home of the vivid sea lettuce, formed of ruffled or flat segments about the size and color of lettuce leaves. Clumps of dense, mossy green algae also grow near the shore. Brown algae blanket the middle- and low-tide zones because they have the resiliency to withstand the full force of the sea. However, the many-branched brown rockweed of the upper tidal zone has adapted so well to living in air that it cannot tolerate continual submersion.

Green chlorophyll in brown algae is masked by yellow and brown pigments which help utilize the limited spectrum of sunlight in depths up to sixty feet. Pigment of the red algae absorbs and uses the dim blue and green light that penetrates farther, allowing members of this group to live at depths as great as six hundred feet. This may be why the red algae are the most abundant seaweeds of the world, numbering four thousand species, and live in habitats ranging from the highest intertidal levels to the great depths just mentioned.

Coralline red algae cover sections of this reef like paint, with a light pinkish-purple film. Jointed calcareous algae, two or three inches long, with similar coloring, are also common here. Small uncalcified pads between the hard segments provide flexibility.

Red algae are washed ashore in great number and variety by the first heavy winter storm. This is an ideal time for observing, photographing, or collecting. These lacelike plants are quite easily dried, pressed, and mounted to retain both color and form.

I see growing with the rockweed the ruffled purple blade of the edible seaweed called nori. In Japan almost half a million workers cultivate nori, which is grown on nets stretched over the length of shallow bays. After harvesting, the nori is chopped in fragments that are spread on mats to dry as thin sheets. Two and one half billion of these nori sheets are produced and consumed each year.

Most marine algae reproduce alternately through sperm-egg union and through the production of spores. A generation of plants derived from spore-producing seaweeds grows to be male and female. The sexually produced generation that follows will grow spore-bearing plants only. Both sperm and egg can be observed only with the aid of a microscope.

A streamer of spray from the first wave to break over the reef abruptly halts my camera work. I move shoreward dripping salt water from nose and chin, leaving the moving patterns of glistening oarweed and bubbles. The sting and bitter flavor of flying spume lift my spirits. I remember as a child the shock of my first taste of sea water. "Why is it salty?" I probably sputtered. To answer this question completely would be to know the action of rain on every slope of the land, the course of all the streams and rivers, the energy of the sun warming the oceans, and the flow of winds and currents. While it is obvious that the ocean is salty because it contains dissolved minerals, the question of where the salts came from is much more complex.

Every year the oceans of the world receive three billion metric tons of material that is drained from land by rivers or blown from land by winds. This includes volcanic lava and ashes, atmospheric dust, and meteors from space. Billions of gallons of water are evaporated from the surface of the sea every day, increasing the concentration of dissolved materials. Living organisms in the sea absorb and use chemicals once held in solution. After an organism's death, these chemicals are released to be used again in the seas' recycling. When small plants and animals die, their bodies sink toward the bottom. Those that are not eaten by creatures living below dissolve completely or become part of a growing layer of sediment on the bottom. A small, single-celled plant like a diatom would take many days to sink one mile. Oceanographers estimate that the dissolved organic material in the sea is equal to about three hundred times the amount of living organic matter.

Despite the vast quantities of mineral and organic materials being dumped into the oceans, the composition of sea water has not changed since measurements were first made. If this constant addition of chemicals were cumulative, the sea would have reached its present degree of saturation in just a few million years; but there is evidence that the oceans are much older. Salinity remains constant at three and one half per cent with only very small variations in different areas. The stable content of sea water is a vital, almost magical phenomenon. Great quantities of ingredients are constantly being added and used up. Gases are absorbed from the air and used in photosynthesis, which releases oxygen as a by-product. Water flows from land to sea and is evaporated. The ocean cauldron is constantly stirred by currents and winds. Sea water cannot be made by mixing all the ingredients in the laboratory, and a bucket of this dynamic substance, separated from the sea, will support life for only a short time.

A series of complex interactions maintains the mineral balance in the sea. Melting ice and rain added to the discharge of rivers supply fresh water. Salts are lifted from the surface of the sea by flying spray, to be carried by wind to land. On an island the size of England, thirty-six pounds of salt are deposited per acre every year by this process. Most of this salt will be carried by rain back to the sea or to inland lakes. Huge supplies of calcium are taken from the sea by shell- and bone-building organisms. Silica is carried to the oceans by river water, which contains five hundred times as much silica as the sea.

Plain water in liquid form, even without any ingredients of the sea soup, seems especially formulated to support life. If it did not occur as a liquid within the temperature range of this planet, world history would be very different. Water can dissolve more substances than any other liquid. It makes up a large part of all living organisms and participates in most chemical reactions involving life processes. Our human bodies are seventy per cent water.

The sea holds at least traces of all the natural elements. One cubic mile of ocean water contains an average of one hundred and sixty-six million tons of dissolved salts. There is enough salt in the sea to cover all the land to a depth of five hundred feet. Common salt, sodium chloride, makes up about 77.8 per cent of all the material dissolved in sea water, magnesium chloride about 10.9 per cent. Magnesium sulphate, calcium sulphate, and potassium sulphate together account for 10.8 per cent. Traces of the other elements make up the remaining .5 per cent. Some of these trace elements are found in much higher concentrations in certain marine plants and animals. The iodine in seaweed is one example.

These salts in solution add weight to sea water. When evaporation on the surface increases the percentage of salts, the heavier water tends to sink, creating vertical circulation. Changes in temperature also cause changes in the density and weight of water that stimulate vertical movement. Surface water of the Mediterranean Sea is saltier and heavier than water of the Atlantic Ocean because of increased evaporation. This sinking heavy water finds an outlet at the bottom of the Strait of Gibraltar, where it flows over the continental shelf to form a mighty subsurface river running south all the way around the Cape of Good Hope at the tip of the African continent. The circular planetary currents, tides, winds, and storms also contribute to circulation.

Sea salt is a by-product of massive interactions between land and sea that circulate materials into the oceans and back to land. The earth's crust moves to compensate for its losses from drainage of materials into the sea. Undissolved debris settles to the bottom, where it joins remnants of organic life to build up sediments forming rock under water pressure. As sections of the earth rise, these sedimentary layers become subject to erosion and may again drain back to the sea.

If the cycles of rising and sinking landmasses had never occurred and the earth's crust had remained fixed when the ocean basins were first filled, the rains, the ice, and the winds of several billion years would have helped the sea to conquer the earth completely. The oceans would cover the whole planet to an average depth of twelve thousand feet.

Statistics only hint at the vastness of the sea; the ocean now has twice the area of land, the volume of water in the oceans is eighteen times more than the volume of land above sea level. Mt. Everest could fit into the thirty-five-thousand-eight-hundred-foot-deep Mariana Trench leaving a mile of water between the tip of the mountain and the surface of the western Pacific.

Here at the base of the cliff, the water-sculptured reef is covered by eroding blocks torn from the land wall. Formed of purple and gray pebbles, this beach is my favorite place for recording the varied patterns of erosion. I find new images here on every visit. Each tide brings in a fresh supply of pebbles, shells, and seaweeds and rearranges familiar objects. Today as I move slowly up the beach from the submerged reef, looking for treasures among cast-offs of this morning's storm, I find black tar plastered over the seaward surface of rock just below high-water mark. Beneath the oil I see the clear concave outlines of three limpets, upturned in death. Part of an abalone shell and one white feather have also been captured by this oily predator. How can I resist recording this negative beachscape? At the same level along the beach other rocks are fouled by wave-borne oil carried in from a very small shipboard oil spill. Plastic debris is so common now on beaches everywhere that I have almost learned to block it from my vision. But the presence of oil on this beach that is so special for me is particularly disheartening.

This small patch of oil that only killed three limpets is part of more than five million tons of petroleum wastes that are dumped in the oceans each year by tankers, offshore industries, municipal oil users, and oil-spilling motorboat engines, according to a document prepared by the National Academy of Sciences. Throughout the world, observers of oceans and beaches are now quite certain to find this unwelcome ingredient in the sea soup. Thor

Heyerdahl, drifting across the Atlantic in a boat of papyrus rushes from Morocco to Barbados Island in 1970, radioed from the mid-Atlantic that oil polluted the ocean from horizon to horizon. He and his eight-man crew were reluctant to bathe or brush their teeth in ocean water. On a recent cruise of the Woods Hole research vessel *Chain* to the southern Sargasso Sea, the nets towed to collect surface life collected three times as much tar as sargasso weed.

When eight hundred and forty thousand gallons of fuel oil spilled along the coast where I live, concerned citizens descended to the beaches by the thousands to remove the black ooze from the sands, the rocks, and the tide pools. This was just one of eight thousand oil spills that occur in United States waters each year. With other volunteers I worked with shovel and hoe on a remote beach made accessible by hacking at the cliff with a bulldozer. Tools, clothes, and all extremities were quickly coated with a heavy asphalt-like mixture. Many tons of oil were hauled up the cliff in gunnysacks and plastic bags, but much of it stayed on the uneven rocky shore, in cracks, crevices, and in spaces between large rocks. Perhaps the most futile aspect of the cleanup was our effort to save thousands of shorebirds trapped in the oil. Even though the birds that could be found were carefully cleaned and cared for, of the seven thousand seabirds injured by the oil fewer than ten per cent survived.

A study by marine biologist Gordon Chan of College of Marin, in Northern California, showed a sixty per cent decrease in living marine organisms on one reef affected by this spill. Whereas fifty-one to seventy-five per cent of the rocky shore behind this reef was oil covered, ninety-six per cent of the barnacles were killed by oil contamination. This percentage of casualties was determined by a count made three months after the accident and then compared to a census made a year earlier. Scientists of Woods Hole Oceanographic Institution investigating a seven-hundred-ton spill of fuel oil in Buzzards Bay off Massachusetts discovered that marine organisms were being poisoned by oil eighteen months after the spill. Hydrocarbons from oil enter the marine food chain, where they become concentrated as they pass from prey to predator, increasing the danger to both animals and man.

Oil spills will continue to cause concern, outrage, and the death of birds and other sea life, but oil tankers still risk collisions as they grope through foggy harbors and we, the consumers of petroleum products, will use gasoline to reach the beaches where birds will be trapped by the next spill.

Petroleum is formed when organic materials deposited at the bottom of the sea are covered with thick layers of sediment. After millions of years of great pressure by rock and sea combined with elevated temperatures, an immensely complex mixture of hydrocarbons is developed. The world supply of natural oil is finite while the energy needs of a growing human population seem infinite. Since oil and coal are concentrated energy from the sun, why don't we develop more direct access to solar power?

Caught in the bind between decreasing supply and increasing need, civilized man is frantically expanding his search for oil. Now that technology has given us mastery over all the remote inhospitable regions of the world, oil will soon flow from the Arctic and from more and more of the seas and oceans. Potential oil reserves in the Arctic amount to five hundred and fifty billion barrels, about equal to those of the rest of the world. We will have to wait for a large oil spill on or below ice to know if world temperatures and weather will be changed by oil melting the ice. As tankers double and quadruple tonnage capacity, oil spills will increase in size as well as number.

No economically competitive substitute for the energy in a gallon of gasoline is available or is being developed, but this is only one of the reasons why oil resources are being exploited with little regard to the impact on our environment or to inevitable total depletion. Oil-producing countries and corporations become increasingly wealthy and powerful. Embargoes and temporary and artificial shortages are now successful economic, political, and military weapons.

Nonpolluting energy from sun and wind can be converted for use in many ways, but so far the cost of these alternatives has seemed excessive compared to the cost of power produced by fossil fuels. Industries based on oil production and consumption have understandably shown little interest in developing competitive power sources.

Oil is but one of thousands of toxic materials that industry dumps into rivers and oceans. Just about everything we throw away in liquid form reaches the sea; chemical effluents, heavy metals, dry-cleaning fluids, radioactive wastes, chemical-warfare gases, detergents, and pesticides are a few of the unwelcome additives now part of the sea soup. Much of the waste discharged into the atmosphere, such as wind-borne lead from gasoline, falls into the sea.

Radioactive wastes are being dumped into the Atlantic and can now be detected in samples of water from all the oceans. The United States Atomic Energy Commission has consigned to the sea low-level wastes mixed with concrete and contained in steel barrels. The British discharge one hundred thousand curies of radioactive waste a month from a pipe running two miles into the ocean. Recently a Russian scientist, G. G. Polyharpov, has discovered that even low radiation levels, two-tenths microcurie, are harmful to the development of eggs of some fish species.

Man has always assumed that the vast oceans contain an infinite supply of food and that wastes added to the sea would disappear, like dirt swept under a rug.

We learn slowly of the fine balance between abundance and shortage, between existence and extinction. Agricultural and industrial wastes from the Sacramento River and from San Francisco Bay pass into the sea over the spawning grounds of the dungeness crab. This legendary delicacy of the Northern California coast has almost disappeared although the size of crabs that may be taken is limited by law.

Because more than half the world's population depends on fish for protein, fleets of sea-going fish factories comb the oceans for a share of this diminishing supply of food. Radar has eliminated the guesswork from fishing. Soon satellite photos may help to locate schools of fish.

The fish population of the sea is concentrated over the continental shelves, so productive fishing is limited to a small part of the total ocean area. When fish became scarce in the northwest Atlantic, Canadian and Russian fleets moved south in 1963 to fish over Georges Bank and Browns Bank, off the New England coast. Until 1962 a plentiful supply of haddock, the most popular food fish of the northeast United States, was taken from these waters by American fishermen. In the last ten years the haddock population has dropped from the hundreds of millions to the tens of millions.

An International Commission for Northwest Atlantic Fisheries is operating to conserve the fish supply, but there is no way to enforce conservation control and some fishing countries, such as Japan and East Germany, don't belong to the Commission. World fishing now takes out more than half the maximum sustainable yield of Georges Bank and an even larger percentage of the haddock.

Here at the edge of the vast and powerful sea the realities of depletion and pollution are hard to accept. From cliff and surf, the roar of the rising flood resounds faster and louder, the reef shatters blue-green swells into gleaming white, the cyclic pattern of millions of years moves unchanged, except for that little blob of oil that smothered three limpets on this rock. The total limpet population of one rock is gone. I know how these limpets died and I am uneasy about the future of limpets on all beaches. I am trying to understand why oil is increasingly fouling more beaches.

Wilderness was an enemy to be overcome and used by the pioneers in our land. Then survival depended on individual strength and initiative. Early farmers developed increasingly efficient tools for controlling land. In the process, most of the forest that covered the eastern part of our country was destroyed, but until it was gone it was not missed. Man has tended to simplify plant and animal life by eliminating species that were not directly beneficial to him, thereby destroying some of the interacting checks and balances that stabilize the natural biological community. Machine and chemical aids have been developed to maintain productivity, but almost each new remedy has become a source of new problems.

Man has emerged as a geographical force on the planet. Whole regions have been changed in appearance and function. Thousands of square miles of productive land have been purposefully destroyed by herbicides in Southeast Asia during an undeclared war. The continuing arrogant use of technological power on land is now beginning to change the oceans.

We are told that human knowledge doubles every ten years. In the past, technology was expanded primarily to increase our control of natural forces. We split the atom, proved its destructive power, and now spend vast sums increasing our overkill capacity while showing little concern for the disposal of atomic wastes that will be lethally radioactive for thousands of years. The Atomic Energy Commission advises that there is no threat from radioactive wastes that have leaked from ruptured storage tanks near Hanford, Washington, because the deadly material now in the earth has not yet reached the water table.

The traditional division of the sciences has made it difficult for man to learn more about himself in relation to the rest of the world. Social sciences and the humanities have long been oriented to the experience of man. Economics and ecology still coexist at opposite ends of the hall. However, a small number of observers and thinkers from many disciplines have always been concerned with the impact of man on his environment. These were the first ecologists, studying the relationships between living things and the physical world. They were the leaders of a new band of pioneers who have learned to reject the value of man's selfish exploitation of the world. These pioneers—biologists, oceanographers, artists, housewives, and school children—all understand that human survival depends on our ability to live as part of the natural world and to maintain the diversity of living forms which in turn support us.

Those who look closely at the natural world, naturalists and artists, are always the first to see and identify rips and tears in the web of life. Their findings are often unwelcome. Their proposals are considered hostile by politicians who often measure progress by increases in the gross national product. Environmentalists are now being blamed for inflation and energy shortages. Hard-won conservation projects are being rolled back. We are told that clean air and clean water are luxuries that we cannot afford. As long as men with these values control our lives, the future of all the limpets and all living communities, including our own, is uncertain.

On the long white crescent between surf and cliff, shifting arcs of wet sand are touched by the sparkle of afternoon sun on billions of sand grains. Between the submerged reef and the distant rocky point, a rising tide and the gently sloping beach are midway in their diurnal encounter. Along the westward-facing edge of this land, late afternoon is my favorite time to explore the open reaches where foam, sand, sun, and wind come together. I hear a shoaling wave crest smack into the trough ahead as the moving energy from distant storms is transferred to seething molecules of white water. Then comes the rumble of tumbling foam as breakers push forward against the friction of the rising beach. With sand squishing between my toes and icy foam swirling around my knees, I am off down the beach, rejoicing with each surge upon the sands. I feel free near the sea as it stretches unhindered to meet sky at the horizon. Only in the desert or on the summit of a high peak am I as aware of space above and around me.

Like handfuls of foam blown by the wind, a flight of sanderlings wheels and swoops low against the indigo sea. With a flash of white breasts, then dark backs swirling in perfect rhythm, this shorebird cloud drops as one bird to the wet sand and resumes feeding. Driven by a desperate hunger, the birds follow the receding surf, snatching morsels of food left by the backwash, ignoring the next towering upsurge until the last instant, when they nimbly scamper back. Heavy surf following this morning's storm brings a hearty supply of fresh food to these migrating surf feeders, who prefer sampling the sea soup from the edge of the bowl.

On this spring day the sanderlings, small members of the sandpiper family, are en route to their breeding grounds above the Arctic Circle. These world travelers could have started this flight in Argentina, New Zealand, or Australia. Twice a year they fly as far as eight thousand miles. This refueling stop is important because they will find scant food on the barren rocky tundra which is their destination. After mating, they will build nests on the ground that hold as many as four olive-green or brown eggs. The camouflaging coloration of eggs and birds will be their only protection from predators before the young can fly. Until insect larvae grow in Arctic ponds after the summer thaw, seeds and buds must sustain them.

Sanderlings weigh only two or three ounces, yet stand eight inches high and have a wingspread of up to fifteen inches. A white stripe along rust-colored wings helps identify the sanderling in flight. These three-ounce navigators generate the energy to repeatedly span continents, flying as high as fifteen thousand feet. Their accomplishments seem miraculous. The energetic feeding habits of these birds fuel hot metabolic engines, for without food every few hours they perish from hunger or cold.

As I stride along the surf's edge, the sanderling flock keeps pace with me, black legs and feet a blur of action, and black beaks guided by sharp eyes busily probing the water film. Annoyed by my persistent intrusion, they fly in a half circle, quickly returning to the sand beyond me. After repeating this flight pattern several times as I approach, they finally fly in a wide circle to continue feeding far behind me.

Like water, sand moves constantly with waves, currents, and winds, accommodating a relatively gentle transition of energy from sea to land. Unlike waves in an uncompromising collision against rock, swells charging against a sloping sandy bottom lose momentum and break into surf before reaching shore. As a breaker flattens in its final rush, patches of blue water break the foam behind its leading edge into streamers and bubbly tendrils. Finally a thin, clear sheet of water glides gently up the smooth sand, slowing until the white rim of the surge stops and slides back. In these final moments the uprush pushes a line of sand particles before it, leaving them there as water moves back or sinks into loose sand. This raised line or swash mark records the progress of the wave until it is erased by rising surf. As the tide ebbs, a pattern of swash marks graphs the height of each receding wave. Sometimes fine strands of surfgrass caught in the uprush add a beautiful linear accent to this daily record.

The tumbling roar of surf softens to a sigh as uprush slides over sand. In the quiet instant before backwash begins, a faint pop of bubbles and a whisper of slack water sinking through sand overlay the steady rhythm of beach sounds. Even quieter phrases in this audible harmony are played by streamlets tinkling and clucking through seaweed shards and dripping from shells and driftwood. These gentle melodies are reflected from the overhanging cliff to add texture and richness to the basic beat of sea on land. In a moment the mutter and grumble of retreating backwash in its turbulent clash with the next surge will replace the soft voices of the upper beach.

Water sliding back down the moderate slope of this beach forms an intricate diamond-shaped pattern of shallow valleys in the coarse sand. Water absorbed by sand during the flood migrates to the surface as the tide ebbs to drain down the slope in delta-shaped rivulets known as rill marks. These streams spread outward to form delicate, plantlike designs. The few hours of direct sun today have already dried the surface of the upper beach. Mysterious little holes and domes appear as uprush skims higher over this loose sand. Air displaced by water sinking vertically into the loose-textured beach rises in a stream of bubbles that leave tiny holes. Swashing water from succeeding waves sinks through these holes, making them temporary funnels. Sometimes, when the tide is rising, a thin layer of wet sand will seal the beach surface. When uprush from the following wave sinks into the sand, air is compacted but cannot escape through the thickening layer of water-soaked sand. As more waves increase the pressure, little bubbles are forced together to form air pockets that push domes of sand, half an inch high and several inches across, above the surface. As a child, I imagined that strange creatures pushed up the domes, but they always collapsed with one touch of my finger.

I wonder how many grains I hold in one handful of sand? I can't estimate how long it would take to count even this small number of coarse grains. My fingers are too large and clumsy to hold just one of these tiny rock fragments. Under my eight-power magnifying glass each little rock appears to have a different shape although they are all roughly the same size. On this beach the light-colored grains have a roughly crystalline structure typical of quartz and feldspar, the main components of granite. Each wet crystal is surrounded by its own moisture film. Wet sand is the end product of a long cycle of erosion. It is seemingly indestructible because each grain is cushioned by a layer of water held in place by capillary attraction. This moisture cushion prevents individual grains from rubbing against each other no matter how great the turbulence. The micro-seas and -oceans between sand particles are inhabited by some of the smallest organisms of the drifting plankton. Each minute rock held in suspension by the beat of the surf is a submerged island in a watery universe.

An average-sized sand grain weighs only about two and a half times as much as the same volume of water, but more than two thousand times as much as an equal volume of air. This is why beach sand within reach of the tides is almost always on the move, while only the smallest sand particles are wind-borne. Wind-carried sand grains without a protective moisture covering are more rounded and often have frosted surfaces. Shells and coral, ground in the mill of ebb and flood, are the basic elements of beach sand in some parts of the world. Microscopic plants and animals do not live in the water surrounding these organic particles, possibly because of the presence of an alkaline solution of calcium carbonate leached from these materials.

With the aid of my magnifier I can see that these bright little stones are all quite different from the rocks and pebbles here. I wonder where they come from? What is their age? Sand and sea salt have the same genesis: water, ice, wind, and roots work with gravity to transform every sloped land surface into its basic components. Sand is the undissolved end result of this process, while sea salts are minerals in solution. Glacial ice carving the western slope of this continent could have started this material on its journey to the sea. This handful of sand could be older than that; thousands of millions of years ago some of these grains might have reached the sea and been transported here from a river mouth in the Arctic Circle. Perhaps these quartz crystals were crushed and ground in the rapids of some mountain stream and then deposited by high water on a river bank to be overrun by the rising sea. This sand could have been locked into sediment for an eon or two before being released by a rising continent to continue its journey to the sea.

Where will this handful of sand be in another million years? The California Current, angling against this coast, might carry the sand far south to help form a permanent bar sheltering a lagoon, or some southwest gale might drive the grains landward among the dunes at the north end of this beach. Perhaps parallel currents will drop these rock particles into an undersea canyon to help build more sedimentary rock.

Rock fragments held in suspension by moving water are constantly sorted by waves and currents. Heavier, larger particles are deposited on steep, exposed shores while flat beaches attract finer sand because turbulence decreases as surf moves over long stretches of shallow beach. Silt and river clay settle only on the bottom in the relative calm of lagoons and estuaries. This sorting action begins on cliffs and ridges of the highest peaks. The forces of erosion first remove the softest and smallest rocks. Granite monoliths are just beginning to crack as smaller boulders of the same material reach the sea as sand.

The shifting layer of soft particles underfoot reminds me that the beach has changed in the weeks since my last visit. At the end of winter the beach face was higher and steeper. Rock formations near the reef were exposed and the sand below the high-water line was firm. A fresh layer of sand now covers some of these landmark rocks. The whole beach is building up toward the gentle summer contour. It is impossible to know if one wave carries more sand up the beach than slides back with it. But since more than five waves every minute carry sand up and down the beach face, a small shift in the delicate balance of forces that cause movement soon makes a perceptible change in the shape of the beach. Each grain that is lifted lands in a different place and each wave moves millions of particles.

Summer and winter beach profiles are dramatically different. Summer surf is lower than the waves of winter, incoming swells are not as steep. As waves shoal and break, offshore sand is picked up and carried along by orbiting water molecules. Then, in the final surge up the beach, heavier sand particles are left behind by the backwash of the waves. Larger and more energetic winter waves as they break generate turbulence which keeps sand in suspension. Because these large waves tend to be close to each other and move quickly, water piles up on the beach, an action that is balanced by a seaward movement along the bottom. This bottom current carries sand away from the beach and drops it to form an offshore bar, a typical feature of the winter beach. Sand that formed the wide sun-baked beach of summer winters offshore under the stormy surf.

Other changes, far slower than cycling tides and changing seasons, are moving the beach. The western edge of this landmass is rising while the northeast rim is sinking. All the continents are moving in relation to one another.

Accurate surveys of the ocean basins and measurements of the age and location of matching formations confirm a theory that the earth's crust is a mosaic of plates floating on a hot, semiliquid layer beneath the crust. These plates or rafts, thirty to one hundred miles thick, grind and crush together causing earthquakes and volcanoes. Movement along cracks most often takes place in the ocean basins where the crust is thin. Molten rock pushes up through these cracks to become solid. Plates that bend downward in ocean trenches and slide under opposing plates are consumed in the hot lower layer.

One such deep rift divides the Mid-Atlantic Ridge along a line that exactly coincides with a map of recorded underwater earthquakes. This ridge is part of a submarine range forty thousand miles long, winding around the globe like the seam of a baseball.

The notion that the continents were once joined and have drifted apart occurred to Sir Francis Bacon in 1620 as he pondered over the matching coastline of Africa and South America. Geologists, using radioactive isotope dating systems, have recently compared rock strata on the west coast of Africa and the east coast of South America. The layers compared were exactly the same. A technique that measures the magnetic orientation of rock as it was formed has enabled scientists to observe mirror images of formations on both sides of the Mid-Atlantic Ridge. Core samples taken with a new deep-water drilling system showed that the farther the drilling ship moved away from the rift, the older were the formations. By measuring the distance from the rift to rocks whose age is known, geologists can measure the expansion rate of the ocean floor. These measurements indicate that one hundred and thirty-five million years ago North America, Europe, and Asia were one large continent. These land rafts have been drifting apart quite rapidly in relation to the age of the planet. This handful of sand, almost immortal, is moving on a journey far beyond my imagining.

Spread out along the beach between the high-water line of the spring tide and the present level are cast-offs of the sea: debris of all kinds, treasures perhaps, carried in by storm waves to make beachcombing an adventure. I have never seen a spotless beach, for a fine network of remnants covering every fringe of sand offers evidence of life under the waves and on the ocean's surface. This network, changing with tide and storm, is a constant indicator of unseen lives and happenings.

A spruce tree with roots still attached is a new arrival on the upper beach. Splotches of acorn barnacles on the trunk confirm that it has been at sea for a long time. Plankton, including barnacle larvae, find everything that floats. These traveling colonies of barnacles and shipworms are among the few living things cast upon these sands. Parallel currents must have carried this tree south for hundreds of miles from the coast where spruce forests met the sea. Ships' timbers shaped like part of a fishing boat's keel lie nearby, entwined with rope attached to a circular crab pot torn from its offshore mooring by the storm.

Far up the beach something dark and solid projects from the sand. I can remember that this is not a rock but a ship's engine, partly covered now by swirling foam. Flaking layers ot orange and red rust give this chunk of metal a sculptural quality in the late-afternoon light, but I move on, having photographed it before. Long tapered floats of the giant bull kelp lie in twisted masses. These twelve-foot sections of brown algae are the nucleus of great tangled collections of debris: holdfasts of sea palm and oarweed, bits of red seaweed from deep water, green surfgrass, and an occasional bright orange cork float that marked a line of crab pots.

Uprush from each advancing wave sorts and rearranges small remnants of tideline life along the perimeter of its forward movement: I find skeletons of the little mole crab; the gracefully shaped carapace of the cancer crab, its inside surface colored a soft lavender; mussel and scallop shells; bits of the sea-urchin test; and an occasional sand dollar. The sand-dollar shells are probably carried in from the offshore bar. When alive, this relative of the urchin has a green-purple covering of short hairlike cilia that are used to dig into the soft sand where this animal lives.

Along the swash mark of the last wave, the beautiful little blue fishing boat with the purple sail, Velella, small cousin of the Portuguese man-of-war, has run aground. This jelly-fish carries a triangular sail rising diagonally from a flat oval disk, three or four inches across. Long, stinging tentacles trail far below as this living ship sails the high seas. The shape and angle of the sail can be adjusted. Velella is a combination of different living animals: a colony of inseparable individuals all starting from the same egg. The disk and float are thought to be one animal; the fishing tentacles have their own organs; other reproductive individuals make up the colony.

Moving through the edge of the foam, I watch for a V-shaped ripple in the sand just above the receding backrush. This is the only indication that a buried mole crab is ready to fish, with just eyes and mouth protruding from the sand. After the next wave has rushed up the beach and is gently sliding back over the crab's position, two brush-shaped antennae will extend into this clear film of moving water for just a moment and then withdraw through appendages around the mole crab's mouth that pick up the food from the net. Like a flock of birds, a colony of these little creatures tends to move in unison. I watch carefully for a host of little gray faces, each waving antennae that look like a long beard; mole crabs will not wait to be recognized.

Sand crabs are one of the few animals that can live on sandy beaches exposed to the full force of ocean waves. They have solved the problem of survival in the most difficult tideline habitat. Even the sand of this beach is in constant motion. Only a highly specialized organism can take advantage of the constant wave-borne food supply without being crushed or swept out to sea. The key to survival here is the ability to burrow into wet sand and to eat, breathe, and reproduce protected from the surf.

Mole crabs are not like other crabs in appearance or activity. They are one or two inches in length, with a streamlined shape covered with a hard, sand-colored exoskeleton. This shape is designed to move easily through sand and to withstand wave shock. Mole crabs swim, crawl, and dig backward. They move up and down the beach to preferred feeding positions, using moving water for propulsion. The whole colony will emerge from the sand as a wave sweeps past, carrying them as far as they wish to go. Then they dig in with a rotary motion of tail appendages and disappear almost immediately. They dig with truly miraculous speed and dexterity. This movement up and down the beach may be repeated several times during the rise and fall of surf on the beach.

The good life of fishing for plankton in the surf enjoyed by mole crabs is sometimes cut short by shore-feeding birds, surf fish, or larger shore-feeding crabs. Then these highly adaptable little creatures become an important link in the tideline food chain between clear sea soup and larger meat eaters.

The most energetic beach burrower is the little nocturnal beach hopper sometimes called sand flea. Each day before sun-up the sand flea, resembling a large sow bug with long antennae, digs a tunnel several inches deep as protection against the turbulence of high water and the long bills of shorebirds as well as against the direct heat of the sun. The funnel that the beach hopper excavates in ten minutes is comparable to a hole sixty feet deep dug by a man with his bare hands. This animal engineer uses three pairs of feet as a conveyor and the fourth and fifth sets of legs for bracing and propulsion. Only when the tide ebbs during the hours of darkness does the sand flea emerge to feed upon all manner of organic debris left stranded by receding waves.

The sand flea breathes air and is moving from the sea world to live on land. It has climbed to the upper beach, but is still dependent on the sea for food and a moist, cool refuge from the sun. Unable to swim or breathe in water, it is not yet able to find food and shelter above the beach. This little animal copes bravely with the predicament of its evolutionary transition.

Unseen below shifting sand and busy surf, other beach dwellers are enjoying the food-and-oxygen-laden waters of the flooding tide. The long, narrow razor clam, protected by several inches of sand, feeds and breathes by pumping water down through a siphon tube. The pismo clam has a filter built into its siphon that prevents sand grains from clogging its body. Lugworms are nourished by passing sand through their bodies and extracting nutrients clinging to the grains.

The sub-sand community can depend on extremely stable living conditions. Sand retains just about its own volume of water. In deep sand the temperature and salinity remain constant. Only the shallow top layer is affected by sun, rain, or running fresh water. Most predators have learned that more energy is used in digging than is gained by eating the victim when it is reached.

My progress from the reef has been slow as the moving tide narrows this ribbon of sand against the cliff. Since dawn the earth has circled through almost half its daily flight. The sun stands near the horizon, bridging the sea with an iridescent path. With eyes turned toward sun-flecked surf, I stumble over some obstacle into a thin layer of chilly foam. Backwash reveals a slate-gray rock marbled by white shell marks. These marks are from ancient fossil shells, preserved just as they were when strewn by wave or current over layers of sediment. I feel almost as if the rock had moved to intercept me, as I have never seen this formation before. The sedimentary cliffs and rocks here are brown and contain only pebbles. This rock with its fossil records may well be a fragment from a hidden formation protruding where the floor of the sea slopes down to the continental shelf. Its rounded contours indicate endless churning in the surf mill. Here is a treasure to be photographed and long remembered. The shell outlines are perfectly composed and resemble the snail, clam, and scallop homes scattered in profusion along this strip of sand. I am tempted to keep my find although it is a bit heavy, but I would rather remember it here, gleaming in its superb natural setting. To my eye all objects from shore and sea lose their natural luster and beauty when moved to become part of our man-made clutter.

Fossil records of natural history represent a series of miracles. First a plant or animal had to be preserved from normal rapid breakdown until its impression was made permanent in rock or sediment. Then it had to become available to discovery on or near the surface. Finally the ancient message had to be decoded by geologist, zoologist, or paleobotanist.

The opening page in fossil history is dated about two and a half billion years ago. These early fossils were bacteria, microscopic water plants. For the next two billion years life flourished, but most of the records were destroyed or have not yet been discovered. Impressions of life appeared again in rocks of the Cambrian period, about five hundred million years ago. By then the ancestors of all the animal and plant families of our tideline were testing survival skills on ancient beaches. A few of the descendants of creatures of the Cambrian time live now, almost unchanged. These are the exceptions. Some early forms have disappeared while others have changed during evolution to meet life's realities more successfully.

Fossil history seems to confirm that life began with simple organisms living in easy conditions, gradually moving and adapting to survival in more difficult situations. Life evolves its own successful forms and is not easily pushed from well-worn paths. Organisms that became highly complex or very specialized dropped from the record kept in rock.

Although man appeared very recently on the evolutionary scene, his intricate civilization is expanding at a rate that will soon outgrow the finite resources of this world. All life on the planet is caught in the web of his technical knowledge. It is difficult but vitally important that we relate the human experience to geologic time and fossil history.

A clear overview is hard to reach by comparing thousands of years to millions or billions. For many of us, the word "billion" is imprecise, meaning only more of anything than we can imagine. But we have some feeling for the passage of time when expressed in days and years. If the earth's age, which science generally accepts as four and a half to five billion years, were one year, each month would represent four hundred million years, each day would span thirteen million years, each hour five hundred thousand years, and each minute nine thousand years.

So, in March of this year spanning earth's history, the ocean basins were filled and the primeval seas came into existence. The record is too dim for precise determination of the beginning of life. This immense journey started between April and June. Just as soon as the first autotrophic bacteria learned to use the energy of the sun to make food and oxygen, more complex forms emerged. The first clear fossil records show the seas in mid-November teaming with plants and animals. On this time scale life moved from the sea to land in the first week of December. In another ten days the first mammals and the dinosaurs lived. Coal- and oil-forming plants had been growing for several days. The primates developed on the afternoon of December 26, and the hominids appeared shortly before midnight four days later. Man himself arrived about ten p.m. on New Year's Eve.

We have enough evidence to put together a moderately consecutive account of the last thirty or forty thousand years or four and a half minutes. Early man planted his first seeds just before 11:59 p.m. and the Christian era began a few seconds before midnight, December 31. Humans have pumped oil from the earth for only a fraction of a second. At the current rate of consumption, this natural resource will be gone before you can blink.

If I belonged to a different intelligent species and could evaluate the human record objectively, I would marvel at the accomplishments of man, the mammal with a brain capable of original and creative thought. But I might expect him to disappear from the ancient pathway of life in the early hours of the following year according to the time scale described above. He would go the way of other complex and highly specialized creatures, but he would be the first to leave because of an overspecialized brain.

As a human, I refuse to write off man as an evolutionary experiment that failed, in spite of the evidence supporting this view. Our transition from tropical jungle to concrete jungle has been perhaps too swift. The discovery of knowledge and the growth of technology have been so absorbing that there has been insufficient time or effort given to evaluate the consequences. Now that the proportion of hungry people in the world is steadily growing, and the fabric of our overtaxed environment is beginning to pull apart, survival pressure is forcing us to redefine progress and growth. This redefining or re-evaluation is our hope.

We have learned that population growth means potential starvation. We have learned that growth in industry means more pollution and more rapid depletion of irreplaceable resources and we have learned that the growth of power by individual nations means that this power will be protected by the threat of total destruction. And yet growth is the way of all life.

Man's short expeditions into space have given us our first overview of this beautiful small planet. National boundaries do not appear on any of the pictures of the globe, but the scars of clear-cutting and forest defoliation are evident. Perhaps this new perspective clearly showing the fragile web of life that sustains us is a clue to the direction our growth must take. Awareness of our total dependence on a worldwide balance of life is growing, but we still lack the will and the discipline to make the changes that will re-establish this balance. The power seekers that control human destiny do not understand that natural law is infinitely stronger than the power of man to control his environment. We can no longer afford the luxury of nationalism, bigotry, or group separateness for racial or religious reasons. Such conflicts waste energy and resources and delay the movement toward a balanced system.

Knowledge of wilderness is an essential resource for individual human growth and happiness, for we receive comfort and strength by opening our senses to patterns of natural life. The harmony of plants and animals with each other and with their environment is expressed in color and form. This beauty is an outward expression of the quality of aliveness. The search for beauty is renewing, for it is the search for the meaning of life itself. Awareness of man's bond with nature, which is our hope for survival, comes from a feeling of closeness to all life.

I have lingered in the mellow sunshine of late afternoon near the fossil rock, I feel my body and thoughts beginning to slow down near the end of this day of intense involvement. I have been musing over my relationship to the real world around me, and now the tide, the reality of this real world, is reaching for the cliffs again. I must hurry back along the beach and climb the path to the cliff top before the sky darkens. I marvel at the play of horizontal light on the wrinkled cliff and pause for a moment behind a feathery waterfall dropping from the meadow above. I cannot pass without photographing the fine drops of water bouncing in golden patterns against the pebble-strewn beach.

At the top of the cliff I am greeted by the flowering meadow. Now in the muted glow of twilight this meadow feels comfortable and more familiar to me than the reef that is covered again by tossing surf. The lupine, cliff daisies, and poppies can be observed and photographed throughout their blooming season, while the tide-pool meadows I have seen today are revealed just once or twice a year by a minus tide during daylight hours. I like to think of this cliff-top meadow as part of the tideline, for it is separated from the beach only by the height of the cliff and is sustained by fog and moist breezes from the sea.

Looking straight ahead toward the horizon, my eyes follow a long, golden path bridging the sea, to confront the sun directly. Our sun seems very close when seen near the horizon through the filter of earth's dirty atmosphere. The sun is relatively close to earth, only ninety-three million miles or eight light-minutes distant, compared with the next closest stellar system in our Milky Way galaxy, Promima Centauri, which is more than 4.3 light-years away. A light-year is the distance that light travels in a year at a rate of 186,282 miles a second, about six trillion miles.

According to current astronomical theory, our sun is middle-aged but will continue to generate atomic energy for another five billion years, until the hydrogen used in nuclear fusion is almost spent. Then the outer layer will begin to expand to about a hundred times its size and the light will become a thousand times as bright. By then the earth will be baked at temperatures that will evaporate the oceans. The sun, now known as a red giant, will use up all the available hydrogen in about another hundred million years and finally only a small core will remain.

The earth, too, is just about in the middle of its life-span. Cycles of change seem to be infinite and intertwined in patterns that we can only glimpse. Here at the edge of the sea the effects of great cycles and the smaller rhythms of life can at least be sensed because so much energy is focused in this intense environment.

With the horizon as a reference point, the sun appears to drop quickly in the moments before sunset. Peaking waves cast long purple shadows ahead of luminous crests, patterns of foam-trailing breakers are edged with gold, and cormorants streak past for a last try at fishing before the light fails. The fiery glow of the sea fig behind me fades before the last orange sliver of sun disappears behind the fog-wisped horizon. Trees and rocks gradually lose texture and color.

Day is slipping into night, gently, with none of the turbulence and excitement of dawn. Surf sounds seem hushed and the breeze slackens before the afterglow. A fog bank to the north begins to brighten with magenta and lavender. This last show of color moves around the eastern sky and is absorbed by the foam below. The cormorants are back on offshore rocks just as I saw them at dawn.

The first star brings night close as I turn toward the forest path with a feeling of deep peace.

2

5

6

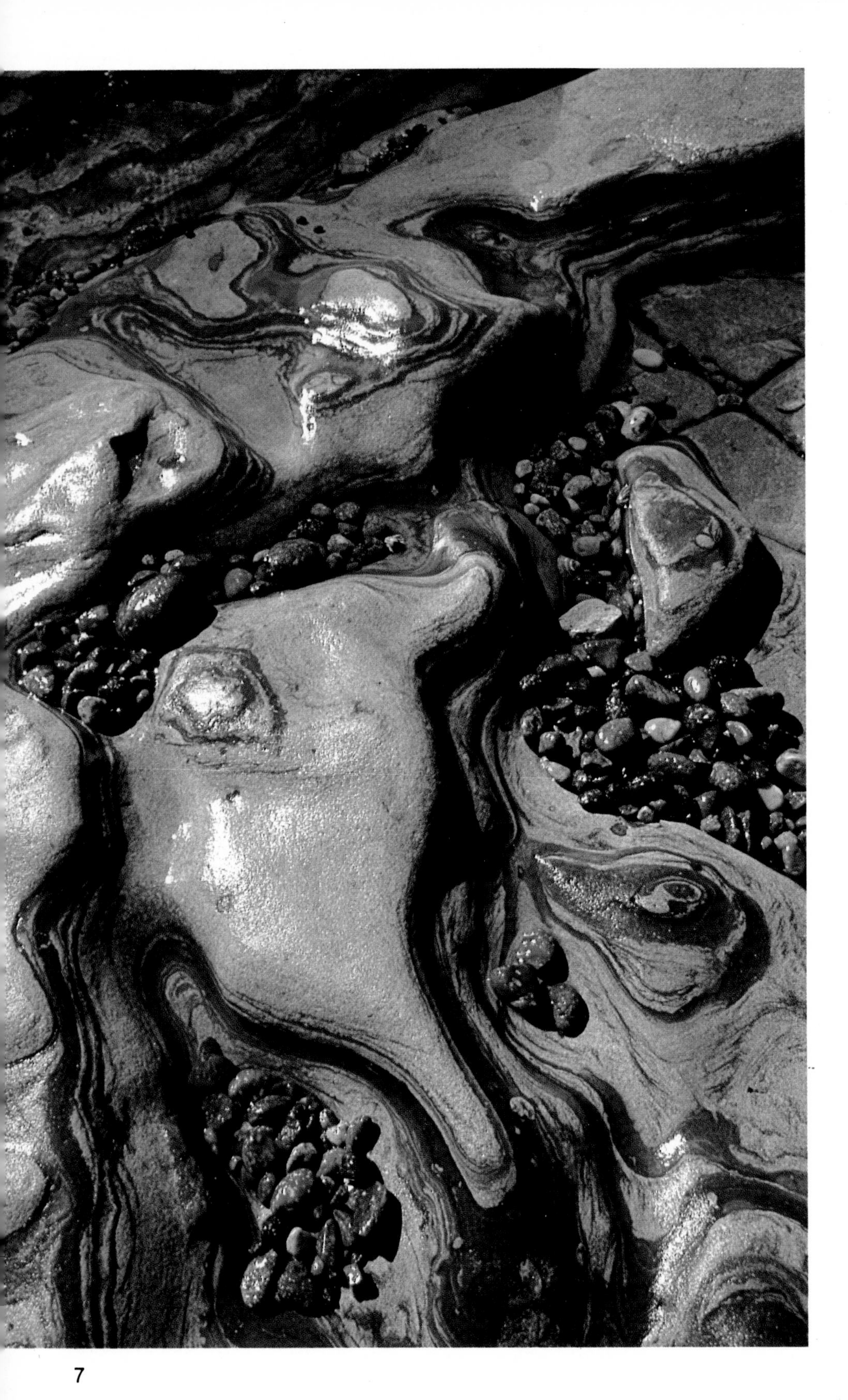

7

8

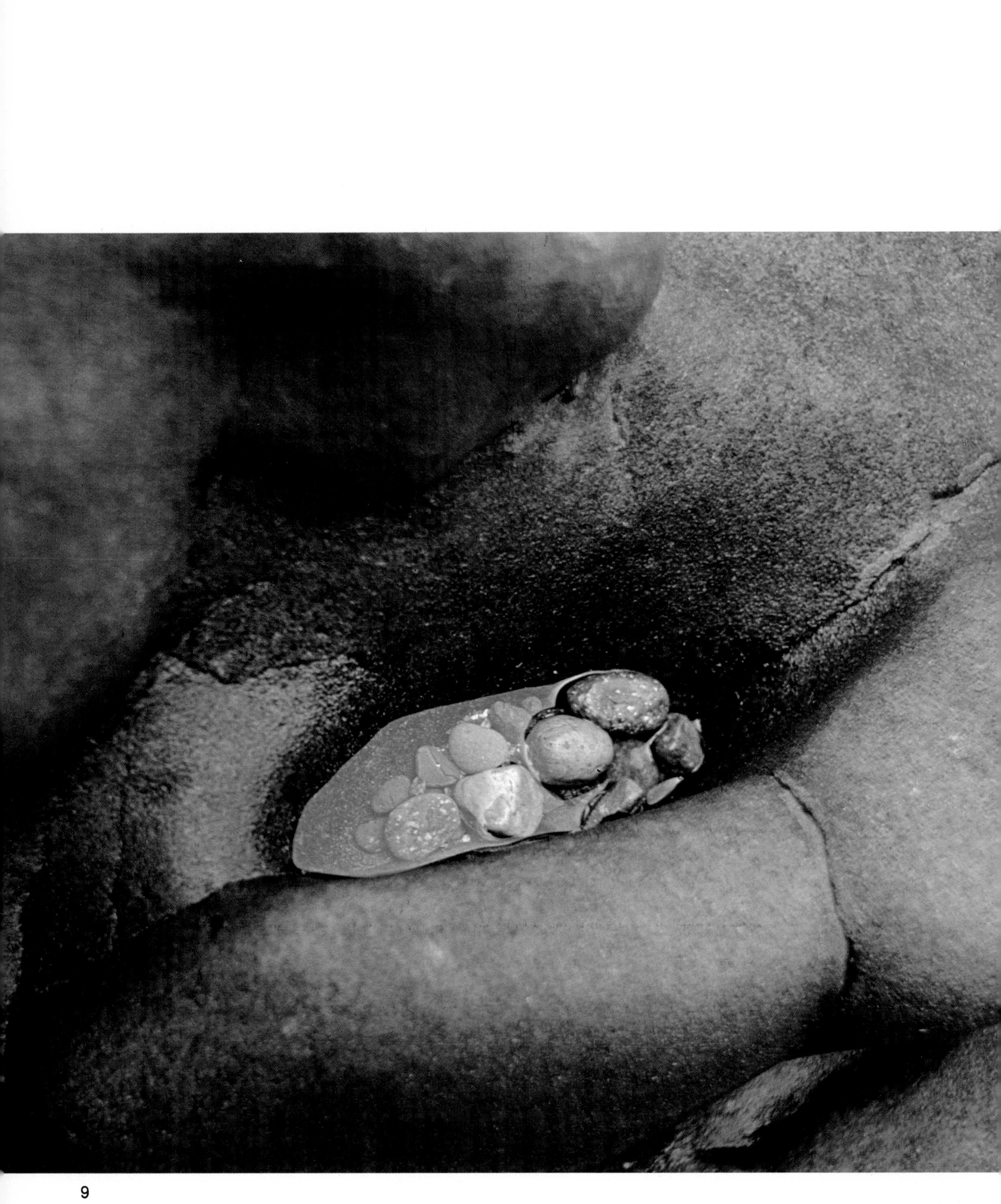

9

10

11

12

14

15

16

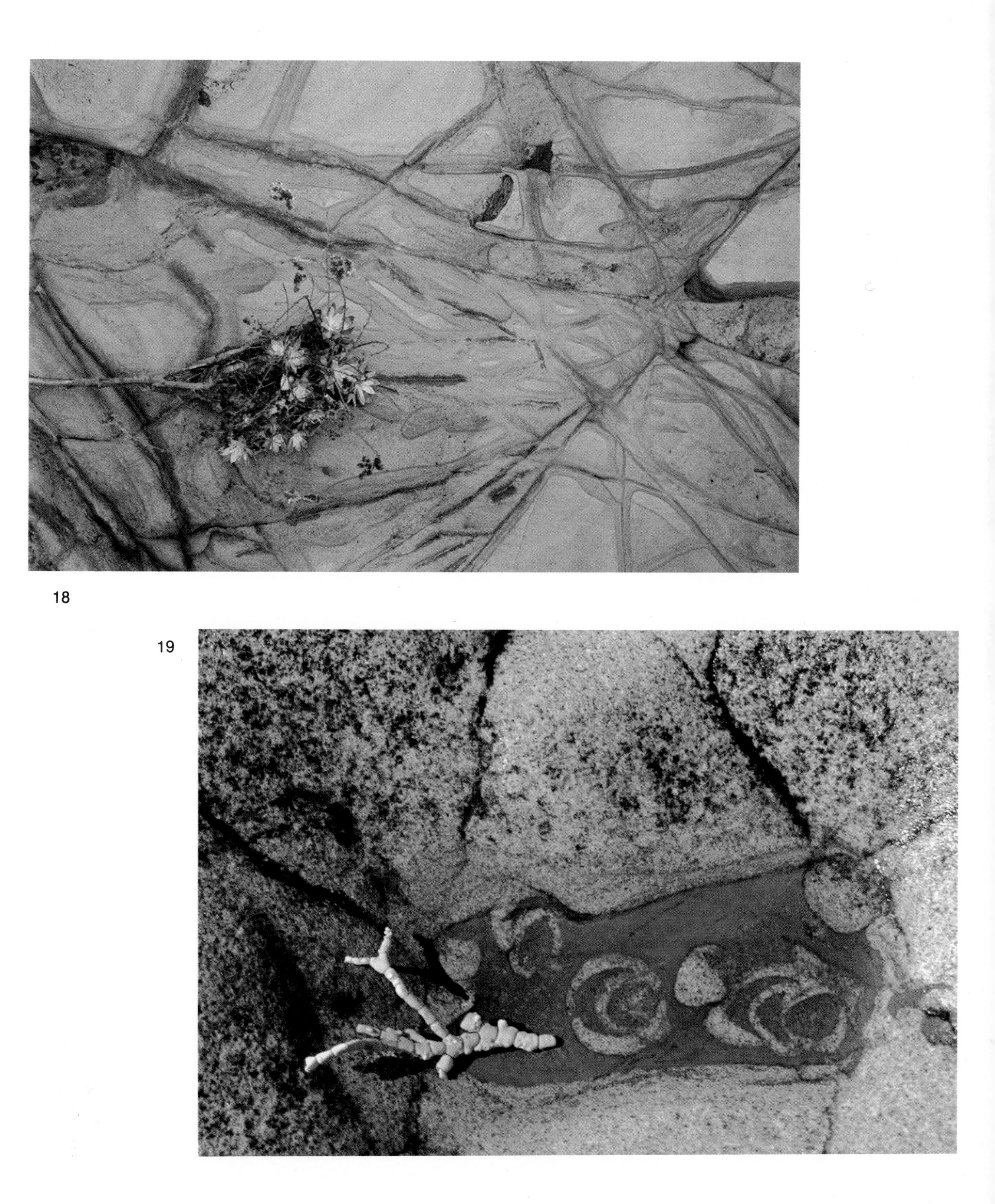

18

19

22

21

24

25

28

29

30

31

32

33

34

35

37

38

39

40

41

42

43

45

46

47

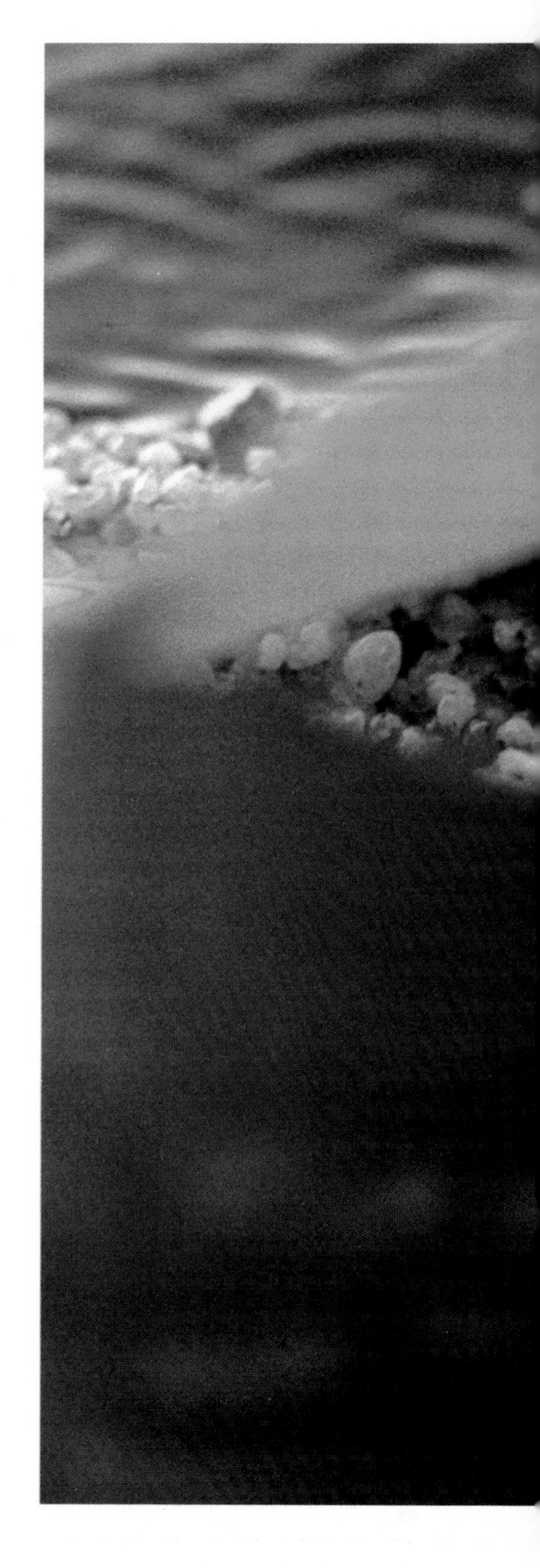

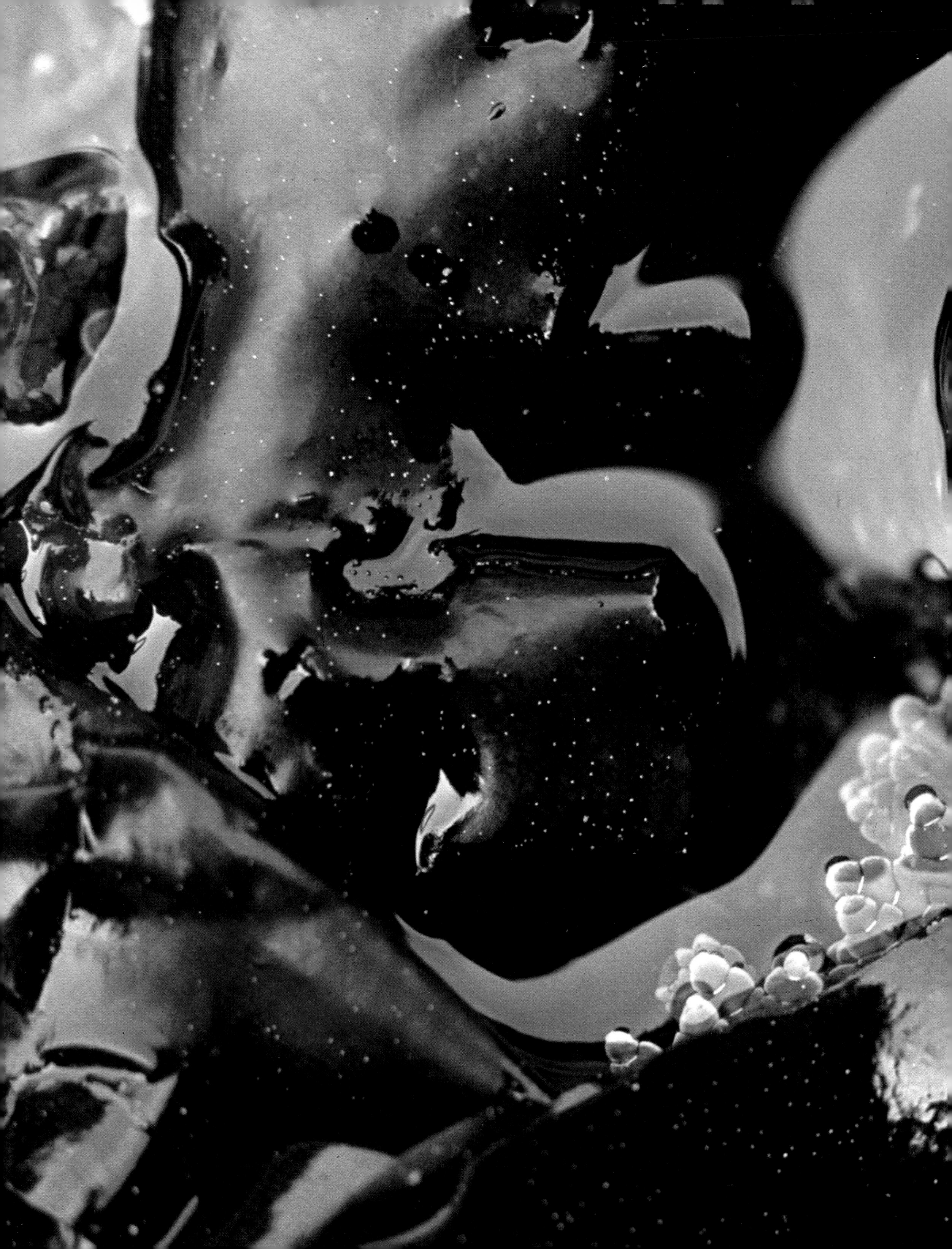

50

51

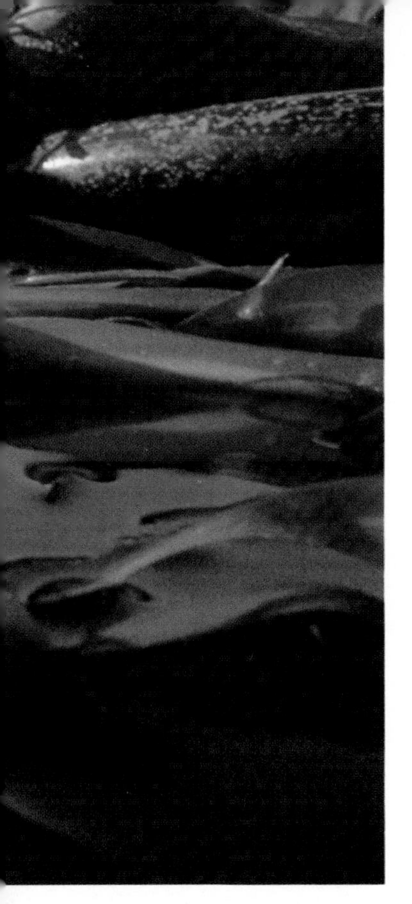

53

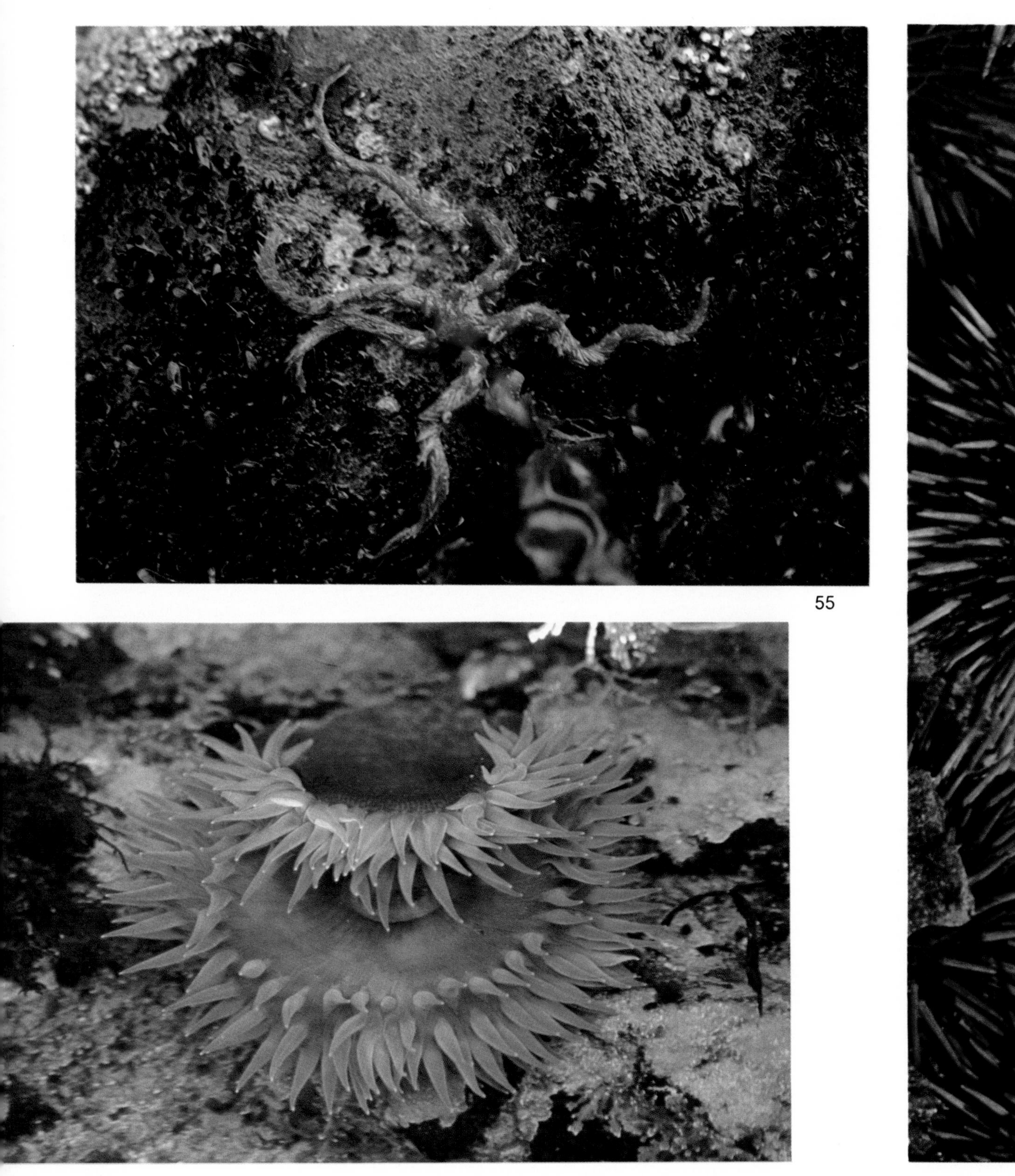
55
56

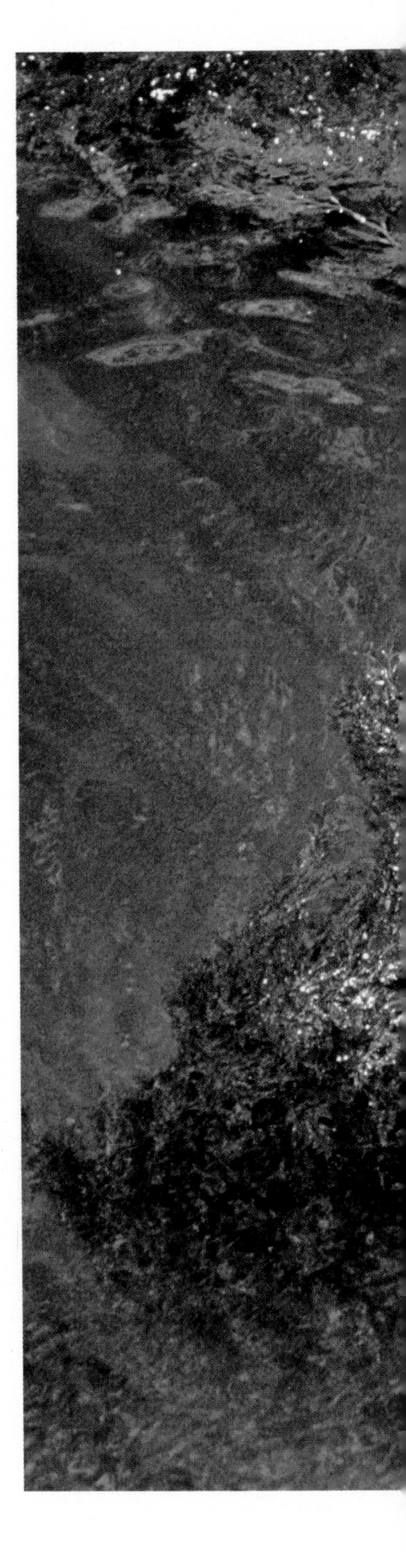

60

61

63

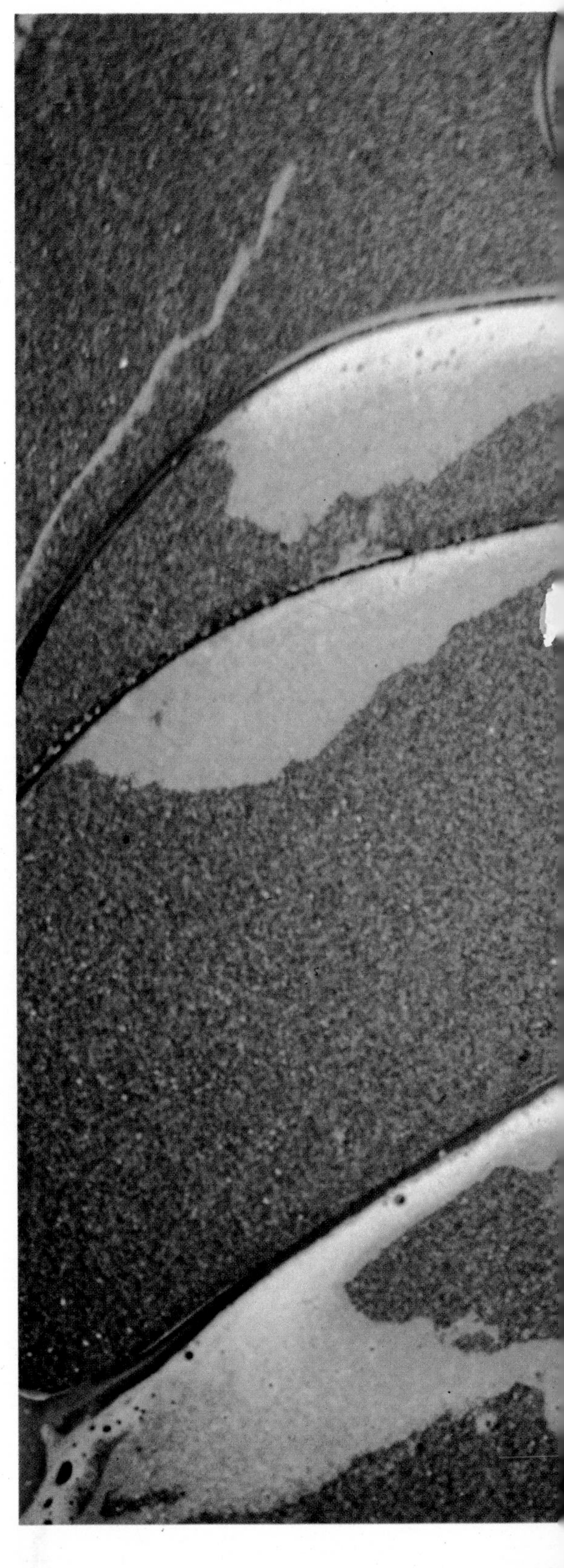

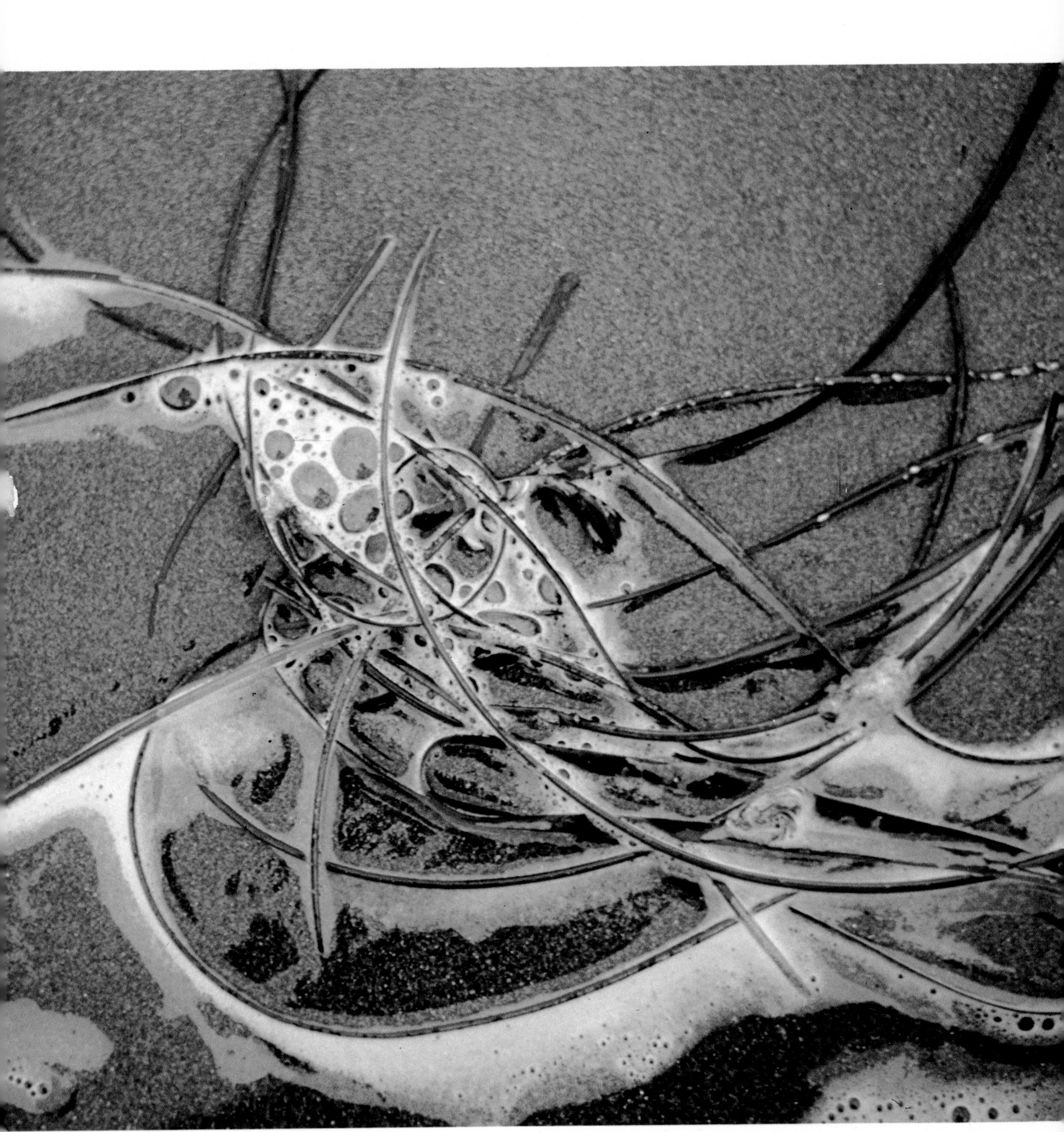

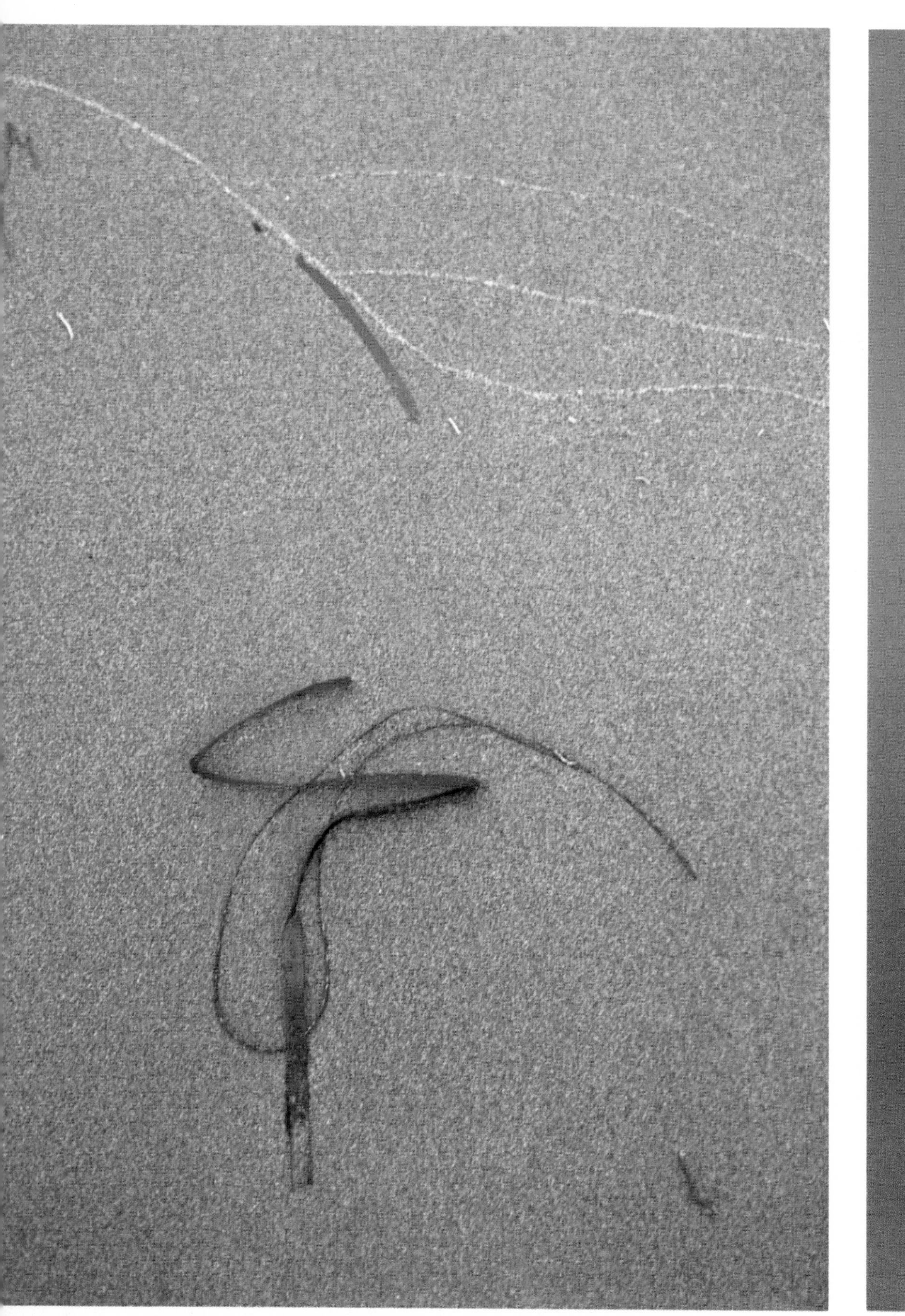

67

68

70

71

73

76

74

75

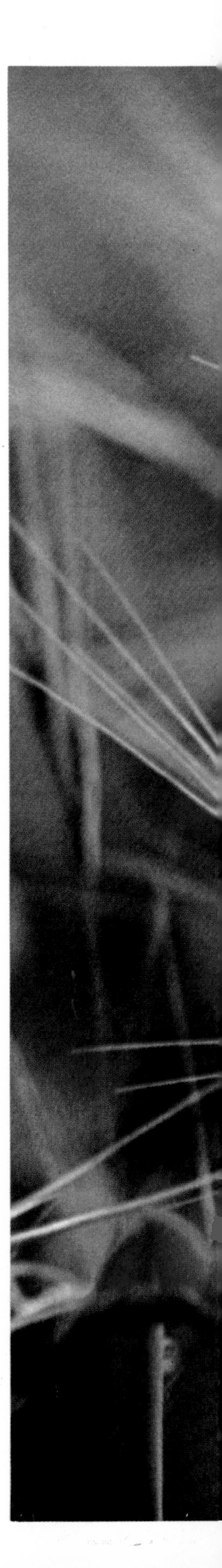

81

Notes on the Plates

1 Morning sun dissipating coastal fog. *200mm lens.*
2 Wave meets rock. *Three-second exposure with 8X neutral-density filter. Kodachrome.*
3 High tide, misty morning. *500mm lens.*
4 Winter surf. *500mm lens.*
5 Breaking wave at dusk. *Six-second exposure. Kodachrome.*
6 Hard pebbles, soft sandstone: the tools and the sculpture.
7 Erosion in sea-formed sandstone.
8 Foam bubbles add a polish to pebbles. *55mm micro-lens.*
9, 10 Water-formed beach sculptures.
11 Pebble-carved abalone shell. *24mm lens.*
12 Winter sun, midday. *21mm Biogen lens.*
13 Offshore wind lifting foam veil from breaker. *300mm lens.*
14 Mossy chiton and limpets.
15 Ocher starfish attacking a mussel colony.
16 Sea soup streaming over gooseneck barnacles and mussels. *One-second exposure. 300mm lens.*
17 Community of limpets at low tide.
18 Cliff lettuce.
19 Tidepool erosion pattern with white coralline algae.
20 Cormorants live on the offshore rocks. *1000mm lens.*
21, 22 The sea fog nourishes cliff-hanging cypress trees.
23 Cliff mosaic of cypress limbs, succulents, and wildflowers.
24 Forest-floor tapestry. *21mm lens.*
25 Poison mushroom pushing up through forest floor.
26 Sea fig and dune grass.
27, 28 Lace lichens flourish in the fog drip.
29 Ground spider webs adorn sword fern.
30 Lichen and pine needles on a tree trunk.
31 Veteran cypresses cling to the cliffs. *85mm lens.*
32 Lace lichen and pine needles.
33 Poison oak, sumac, lichen, and pine needles on the floor of the forest.
34, 35 Forest-floor communities: lichen, mushrooms, grasses, and pine needles.
36, 37 Wild iris at dawn. *55mm micro-lens.*

38 Abalone-shell fragment and seaweed holdfast nestled among beach pebbles.
39 Gooseneck barnacles in the upper tidal zone.
40 Ribbed limpets are upper-tidal-zone neighbors.
41, 42, 43 Feathers on the beach. *55mm micro-lens.*
44 Western gulls. *300mm lens.*
45, 46 Western gulls at rest and lifting off for river-mouth fishing.
47 Small jellyfish stranded on seaweed.
48 Kelp shards tossed up on sand. *55mm micro-lens.*
49 Seaweeds floating in tide pool.
50 Floating seaweeds, low tide.
51 Blades of giant bladder kelp with bat star.
52 Starfish using tube feet to open mussel shell.
53 Surfgrass and oarweed.
54 Surfgrass and foam bubbles. *55 mm micro-lens.*
55 A member of the tide-pool community, the brittle star.
56 Another tide-pool resident, the giant green sea anemone.
57 In the tide-pool community, a crab moves over purple sea urchins.
58 Seaweeds of the tide-pool community: sea lettuce with Gastroclonium.
59 Swirling giant kelp.
60 Foam bubbles on a kelp blade.
61 Foam bubbles in a crab's carapace.
62 Sea shell fossils imprisoned in a beach rock.
63 Foam left by the backrush of the waves marks the tideline. *28mm lens.*
64 Surfgrass and foam on the sandy beach.
65 Pattern left by a receding wave and surfgrass.
66 One blade of grass and one seed: basic elements of life on land.
67 Beach poppy and sea fig.
68 Sea fig and wildflowers.
69 Cliff-top grasses. *85mm lens.*
70 Wild mustard of the cliff-top meadow overhangs sand and surf.
71 Flowers of the cliff-top meadow: seaside daisy and paintbrush.
72 More flowers of the cliff top: California poppy and lupine.
73, 74, 75 Minutiae in the cliff-top meadow.
76 A tide-pool citizen, the ostrich-plume hydroid. *55mm micro-lens.*
77 Poppy petal and filaree blossoms. *55mm micro-lens.*
78 Wind-blown meadow grass. *200mm lens with extension tubes.*
79 Sea fig blossom.
80 Grasses and sea fig.
81 A damselfly rests on a lupine blossom. *55mm micro-lens.*
82 Shooting star or wild cyclamen. *135mm lens with extension tubes.*

83 At dusk, cormorants skim the heavy surf. *500mm lens.*
84 A great blue heron broods on offshore kelp beds.
85 Sunset reflections on wet sand.
86, 87 Sun-glow illuminates a cliffside waterfall.
88 Sanderlings feeding at dusk. *300mm lens.*
89 Sunset afterglow on the beach.

Tideline Photography

As I start to photograph, a strong southerly gale lashes the beach and the offshore wind blasts peaking waves into streamers of foam. I have waited months for the chance to work in the strong wind that just precedes a severe winter storm. The surf is wilder that I have ever seen it. All the joys and problems of outdoor photography face me: the challenge to interpret rapidly changing patterns of sunlight, clouds, water, and rock; the concentration required to retain images of the intense beauty of each passing moment; and the capacity to see clearly and feel deeply the basic reality of aliveness.

Leaning into the wind, I move along the cliffs with camera gear more or less protected by my flapping poncho. In the lee of a rock, on a shelf just above the surf, I finally find just enough protection to photograph without being blown away. I am able to wipe the spray from lenses and reload cameras under the poncho. As storm clouds darken the sea, I crouch by the rock, watching, waiting, and shooting, marveling at the power of wind and water and the exuberant flight of gulls and cormorants close to the rising surf. I continue photographing during the first spattering rain, but the driving downpour that follows soon makes my work impossible. Although cameras function for a while when wet, eventually the moisture inside the eyepiece interferes with seeing. Without the incentive of photography I would quickly seek shelter and miss a great experience.

This adventure, so fresh in my mind today, will not be complete until the film is processed and viewed on the light table. I will eagerly look for pictures that convey the feeling of this storm. If one image of the dozens of exposures made truly documents the vitality of the storm, I will be more than satisfied. An essential part of photographic work is the editing and discarding of all but the strongest images.

Besides communicating simply and directly, a completely alive photograph has a quality of its own. All the elements: line, form, color, texture, combine to create an entity independent of the subject. Such an image is the result of concentration, patience, experience, intuition, and luck. There are no shortcuts to the "alive image" for the photographer or the painter.

The intensity of the living process draws me to the tideline again and again. Every niche is occupied by plants and animals whose tenacity and determination to survive is made clear by their diverse life-support systems. The search for beauty first led me to look closely below the surface of a tide pool. Beauty is an indication not only of the harmonious growth and interaction of animate things in a balanced biome, but something more, a mysterious outward manifestation of aliveness. The quest for beauty leads me through cycles of curiosity, discovery, and sometimes added awareness.

At best the photographic image can convey an essential truth. The image and its message are all-important to me. Technique and equipment are only the means to this end. No two-dimensional image can be an accurate recording of what the eyes see. Color films are never accurate. Black-and-white images are simplifications of reality. A completely sharp photograph is an abstraction, for the human eye scans, seeing clearly just one section of a scene at a time. We gain freedom by knowing that the visual language is always an interpretation of reality.

The technology of photography has developed tools that have expanded our image-making capabilities. The single-lens reflex 35mm camera, interchangeable lenses of many focal lengths, and moderately reliable color film have reduced the mechanical barrier between seeing and making images. I feel closer to my environment when seeing directly through the optics of the camera at the instant an image is recorded. Whatever I see I can photograph. All the pictures in this book were made with Nikon SLR cameras and lenses.

Two cameras, seven or eight lenses, protected by soft leather pouches, fit easily into my day-pack. Because beaches are at sea level and close to transportation, the discipline of backpacking camera gear into wilderness along with life-support essentials is unnecessary. I enjoy the luxury of being able to carry all the equipment that might be useful. I usually work with one camera loaded with Kodachrome 25, which I prefer to use, and a second camera with faster color film for use with long-focal-length lenses which magnify camera movement. An extra camera is security for me, as I have an acquired distrust of all things mechanical, based on experience. Unfortunately cameras are not designed and built for even intermittent exposure to salt water, rain, mud, and sand.

Different camera systems have different strengths and weaknesses. I have used and respected the large-format view camera for thirty years as a working photographer. However, the lens in a view camera that is placed on the ground is still several inches above a barnacle or limpet. The 2¼-inch square-format camera system is heavy to carry and cumbersome to use in the field, and does not have the built-in controls and adjustments of the view camera. The quality of a print or reproduction from large-format film cannot be equaled, but many of the fleeting moods of the tideline can be captured only with the more versatile 35mm format. The 35mm camera allows me to explore the environment in detail. Using available light, I can make one-to-one life-size images of the minutiae of life.

Viewing the natural world very closely through a camera lens can be a lot more fun than peering through a kaleidoscope. I may be belly down on a wet, cold beach with camera at sand level, to record a tiny delicate hydroid before the next incoming wave moves it and the photographer; but in those few moments I am part of the world of the hydroid that I had never seen before. Small changes in position and focus completely alter the image. My freedom of choice is limited only by the validity of the final photograph.The micro-lenses, which allow close focusing without extension tubes, are convenient, but I like the variety of perspective and feeling obtained when using longer lenses such as the 135mm, 200mm, and 300mm for extreme close-up work. Automatic extension tubes simplify this use of longer lenses.

The soft diffused light from an overcast sky is often preferable for close-up photography. Color seems more intense when relieved of the shadows and contrast of direct sunlight. Above the beach, the ground-level community of grasses, flowers, and insects is best photographed in the quiet air and soft light of dawn, when dew or frost ornament all growing surfaces.

The quality of light on the tideline is almost as changeable as the tides. Complex relationships between air and water temperatures and wind and currents combine with regional weather patterns to form unpredictable micro-climates. A clear, sunny stretch of beach may become completely fog-shrouded in minutes when a small drop in temperature converts moisture in the air to visible droplets. This change of water from blue to gray has a profound effect on the tideline mood. Reflections on water and wet sand during a weather change often have a delicate, elusive quality. Photographs of beach or forest in fog take on an increased dimension of depth. Background objects are lighter in value and their form is simplified.

As photographic tools become more adaptable, we have greater freedom to direct and control the style and content of visual communication. Photographers are more limited by the visualization process than by their equipment. In making a photograph, the basic consideration following choice of subject and acceptance of the quality of light is organization of the image. Just as a sentence can be worded for clarity, a photograph can be composed to place emphasis on the essential characteristics of the subject. Selection of a lens of appropriate focal length and control of the zone of sharpness give flexibility to the arrangement of subject elements within the image frame.

The pictures for *Tideline* were made with lenses varying from the wide-angle 21mm to the narrow-angle 1000mm. I try to use the lens that will make the simplest statement. The wide-angle lenses, besides including more of a scene, can emphasize the perspective of close-ups and permit me to relate subjects close to the lens with their surroundings. The long-focal-length or telephoto lenses allow me to choose distant subjects very selectively. The telephoto lenses also make possible direct relationships between distant objects such as birds and waves.

One of the most valuable tools, particularly for extreme close-up work, is the variable depth of field or zone of sharpness. I wholeheartedly disagree with the F64 group of photographers, who believe that the overall sharpness of a photograph is next to godliness. Aside from the fact that the pupils of our eyes do not have an f/64 setting, I find that an extremely shallow depth of field, which means that the lens aperture is wide open, helps me to make simple, strong images. The out-of-focus or soft part of the picture is not negative space but works in counterpoint to strengthen a direct statement.

The grandeur of the assault of sea against land and the varied patterns and forms of moving water are the basic themes of tideline graphics. The power and explosive violence of the sea can be expressed by freezing the action with fast shutter speeds or by choosing slow speeds that let movement register on film. The literal shape of a wave can be recorded by stopping the action, while a feeling of movement and expended energy can sometimes be better expressed with slower shutter speeds.

My first step in making an action photograph is to study the motion cycle and try to visualize the impression of movement on the film. I must decide when to open and close the shutter as well as which segment of action will best express the feeling and mood. The background of the motion is very important. A moving image of a breaking wave will disappear with light sky behind it, but can be seen clearly against a dark background such as a cliff in shadow. Exposure time depends on the intensity of action and the desired effect. I have sucessfully used shutter speeds of one eighth of a second up to six seconds. Of course a tripod must be used for these slow exposures. Neutral-density filters that decrease the amount of light reaching the film are needed in bright light to prevent overexposure, even with slow color films. The results of slow-motion photography are unpredictable, surprising, often disappointing, and sometimes superb.

I will return to the tideline just for the joy of being there; to see, hear, smell, and feel life at the edge of the sea. I will continue to explore the changing patterns of land, sea, and air with a camera. There is reward enough in the work itself as well as the possibility of finding and sharing some essential reality of existence on our small planet.

Additional Reading

Bascom, Willard. *Waves and Beaches: The Dynamics of the Ocean Surface*. Garden City, N.Y.: Doubleday & Co., Inc., 1964.

Bates, Marston. *The Forest and the Sea*. New York: Random House, Inc., 1960.

Carson, Rachel. *The Edge of the Sea*. Boston: Houghton Mifflin Co., 1955.

———. *The Sea Around Us*, rev. ed. New York: Oxford University Press, Inc., 1961.

———. *Under the Sea Wind: A Naturalist's Picture of Ocean Life*. New York: Oxford University Press, Inc., 1952.

Collins, Henry H., Jr., ed. *Bent's Life Histories of North American Birds*, Vol. I, *Water Birds*. New York: Harper & Brothers, 1960.

Holling, Holling C. *Pagoo*. Boston: Houghton Mifflin Co., 1957.

Johnson, Myrtle E., and Snook, Harry J. *Seashore Animals of the Pacific Coast*. Gloucester, Md.: Peter Smith Publisher, Inc.

Kelley, Don G. *Edge of a Continent: The Pacific Coast from Alaska to Baja*. Palo Alto, Ca.: American West Publishing Co., 1971.

MacGinitie, G. E. and N. *Natural History of Marine Animals*. New York: McGraw-Hill Book Co., 1949.

Marx, Wesley. *The Frail Ocean*. New York: Ballantine Books, Inc., 1969.

Matthiessen, Peter. *Shorebirds of North America*. New York: The Viking Press, Inc., 1967.

Peattie, Donald C. *Flowering Earth*. New York: The Viking Press, Inc., 1961.

Ricketts, Edward F., and Calvin, Jack. *Between Pacific Tides*, 4th ed. rev. by Joel W. Hedgpeth. Stanford, Ca.: Stanford University Press, 1968.

Tinbergen, Niko. *Herring Gull's World*, rev. ed. New York: Basic Books, Inc., 1974.

Wayburn, Peggy. *Edge of Life: The World of the Estuary*, new ed. San Francisco: Sierra Club Books, 1972.